Aeronautical Telecommunications Network

Advances, Challenges, and Modeling

Aeronautical Telecommunications Network

Advances, Challenges, and Modeling

Edited by
Sarhan M. Musa
Prairie View A&M University, Houston, Texas, USA

Zhijun Wu
Civil Aviation University of China, Tianjin, China

CRC Press
Taylor & Francis Group
Boca Raton London New York

CRC Press is an imprint of the
Taylor & Francis Group, an **informa** business

CRC Press
Taylor & Francis Group
6000 Broken Sound Parkway NW, Suite 300
Boca Raton, FL 33487-2742

First issued in paperback 2017

© 2016 by Taylor & Francis Group, LLC
CRC Press is an imprint of Taylor & Francis Group, an Informa business

No claim to original U.S. Government works

ISBN-13: 978-1-4987-0504-2 (hbk)
ISBN-13: 978-1-138-74752-4 (pbk)

Visit the Taylor & Francis Web site at
http://www.taylorandfrancis.com

and the CRC Press Web site at
http://www.crcpress.com

Dedicated to our families

Contents

Preface

There are a lot of concerns today regarding the current transportation system growth, procedures, and technologies in aeronautical telecommunications due to safety, flexibility, and growing of new demands. Indeed, the recent growth in aeronautical telecommunication network (ATN) technology has changed the landscape and the role of different systems that give aircraft a link with the ground while in flight. Increase in air traffic transportation increases the communication rate of airborne aircraft for traffic coordination with the ground traffic controller.

It also gives the readers the opportunity to broaden and deepen their knowledge in the advances, challenges, and modeling of ATN. Indeed, it covers recent and future methods and problems illustrated in the field of ATN. It also provides strategies for integrating existing and future data communications networks into a single internetwork serving the aeronautical industry.

The book covers mobile Internet protocol version 6 (MIPv6) as a core of mobility management mechanism for aeronautical environment. The book deals with the simulated performance of VDL Mode 2. It investigates the air traffic management (ATM) system cyber-attacks since the 9/11 event and the current security situation of the ATM information system related to the safety flight of air transportation.

The book provides an introduction to current trends in ATNs, including tools, techniques, protocols, and architectures; MIPv6; air traffic control (ATC); security of ATM; very-high-frequency (VHF) data link (VDL); aeronautical radio and satellite communications; electromagnetic interference to aeronautical telecommunications; quality of service (QoS)-satisfied ATN routing mechanism speed dynamic environments; service-oriented architecture (SOA)-based ATN transmission control algorithm; and the future directions, opportunities, and challenges of ATN. This book has 10 chapters.

Chapter 1 provides an overview of ATN.

Chapter 2 presents an overview of MIPv6 and its extension that can be used in next-generation ATN systems. The authors begin from the infrastructure of general mobile communication, and then the motivation of MIPv6, MIPv6 protocol, and related triangular routing problem is presented. Due to some defects of the original MIPv6 protocol, four types of extension are introduced, that is, Hierarchical Mobile Internet Protocol version 6 (HMIPv6), FMIPv6, PMIPv6, and network mobility (NEMO). Since NEMO is very suitable for a whole mobile network moving, the authors focus on the optimization of NEMO and its applications.

Chapter 3 introduces modern ATC systems from the perspectives of the user and the developer. The composition of the ATC system is illustrated, and some peripheral equipment such as automatic dependent surveillance–broadcast (ADS–B), secondary surveillance radar, and surface movement radar are also provided. Some design experiences are included in this chapter, including the considerations when selecting a better solution. All subsystems of the ATC system are described in the order dataflow. In addition, the authors show how to design an ATC system and how to use it to control in-flight airplanes.

Chapter 4 reviews the security of VHF data link in ATM. It analyzes some main security threats for the Aircraft Communications Addressing and Reporting System (ACARS) data link from the information perspective and offers a detailed solution to varieties of threats.

Meanwhile, it introduces varieties of algorithms used in the solution, such as encryption, authentication, key exchange, and coding and decoding.

Chapter 5 expounds the application prospect of the VDL2 system and simulation. The author provides the VDL technical advantage over the current ACARS and the necessity to realize the system transition.

Chapter 6 presents the special case of Global Navigation Satellite System (GNSS) multipath interference in urban canyon environments starting with background information and a discussion about the evolution of GPS and its interoperability with the Galileo system under the conditions of continued modernization. Next, the author presents the results of his own research to a spatial mitigation technique that capitalizes on the multiple GNSS systems to cut down on the negative impact of multipath. It is expected that this treatment will provide the readers a general understanding of the multipath phenomena and its importance to unmanned systems.

Chapter 7 analyzes the electromagnetic interference of aviation communications equipment, introduces the electromagnetic interference effects and response measures of aviation communications equipment, and then analyzes the protective measures of aircraft and ground stations against electromagnetic interference; it can be referenced for electromagnetic interference protection for aeronautical communications equipment.

Chapter 8 analyzes and models the QoS mechanism in ATN. The author addresses the combination of the ATN environment with the IPv6/MIPv6/NEMO.

Chapter 9 provides the analysis of time division multiplexing (TDM) in a satellite aeronautical communications system when input sources are greater than available channels. The analysis of blocking and clipping probabilities for TDMs was successfully achieved, and results of the analysis were generated.

Finally, Chapter 10 introduces an ATN transmission control algorithm based on SOA.

Sarhan M. Musa

Zhijun Wu

Acknowledgments

Our sincere appreciation and gratitude to all the book's contributors. We also acknowledge the outstanding help and support of the team at Taylor & Francis Group/CRC Press in preparing this book, especially from Nora Konopka, Michele Smith, Jonathan Plant, Richard Tressider, Marc Gutierrez, Arlene Kopeloff, and Hayley Ruggieri.

Thanks to Jean Turgeon (JT), Liam Kiely, Marc Randall, Edwin Koehler, Paul Unbehagen, Edwin Koehler, Mike Nelson, Brian Smith, Aaron Eddy, Roger Billings, Bryan Marklin, Jeff Buddington, and Jeff Cox, Avaya, and Mike Bagby, Skynet Cloud Solutions, for their advices and supports.

Thanks to Professor John Burghduff and Professor Mary Jane Ferguson for their support and understanding and for being great friends. Dr. Musa thanks Dr. Mansoora A. Sheikh for caring for his mother's health during this project.

We appreciate the support received from Dr. Kendall T. Harris, dean of college of engineering at Prairie View A&M University.

Finally, the book would never have seen the light of day if not for the constant support, love, and patience of our families.

Editors

Sarhan M. Musa, PhD, is currently an associate professor in the Department of Engineering Technology, Roy G. Perry College of Engineering, at Prairie View A&M University, Texas. He has been director of Prairie View A&M Avaya Networking Academy, Texas, since 2004. Dr. Musa has published more than 100 papers in peer-reviewed journals and conferences. He is a frequently invited speaker, has consulted for multiple organizations nationally and internationally, and has written and edited several books. Dr. Musa is a senior member of the Institute of Electrical and Electronics Engineers (IEEE) and is also an LTD Sprint and a Boeing Welliver fellow.

Zhijun Wu, PhD, currently serves as a professor in the School of Electronics and Information Engineering Technology at Civil Aviation University of China (CAUC), Tianjin, China. He has been director of CNS/ATM Institute, CAUC, since 2006. Professor Wu has published more than 50 papers in peer-reviewed journals and conferences. He is a frequently invited speaker on electronics and information technology, has consulted for multiple organizations nationally and internationally, and has written and edited several books. Professor Wu is a senior member of the Institute of Communication, Tianjin, China, and a Boeing Welliver fellow.

Contributors

Yun-Fei Jia
School of Electronics & Information
   Engineering Technology
Civil Aviation University of China
Tianjin, People's Republic of China

Saeed M. Khan
Department of Engineering Technology
Kansas State University Salina
Salina, Kansas

Douzhe Li
School of Electronic Information Engineering
Tianjin University
Tianjin, People's Republic of China

Zhao Li
School of Electronic and Information
   Engineering
Civil Aviation University of China
Tianjin, People's Republic of China

Gao Lin
School of Science and Technology
Tianjin Economic and Financial University
Tianjin, People's Republic of China

Zhigang Liu
Airport Management College
Guangzhou Civil Aviation College
Guangzhou, People's Republic of China

Sarhan M. Musa
Department of Engineering Technology
Prairie View A&M University
Prairie View, Texas

Zhijun Wu
School of Electronics & Information
   Engineering Technology
Civil Aviation University of China
Tianjin, People's Republic of China

Meng Yue
School of Electronics & Information
   Engineering Technology
Civil Aviation University of China
Tianjin, People's Republic of China

Haitao Zhang
School of Electronic Information
   Engineering
Tianjin University
Tianjin, People's Republic of China

1

Overview of Aeronautical Telecommunication Network

Sarhan M. Musa and Zhijun Wu

CONTENTS

1.1 Brief Introduction to Air Traffic Management

In recent years, network-based, intelligent, automatic schemes are proposed to solve various challenges and problems in the air traffic management (ATM) field [1,2]: this trend will be sustained in the next decades or even in the future. For better understanding, a summary of ATM and its current development will be introduced in this chapter.

Before any description, a clear definition of ATM given by the International Civil Aviation Organization's (ICAO) document DOC4444 [3] should be pointed out, that is, "the dynamic, integrated management of air traffic and airspace including air traffic service, airspace management and air traffic flow management—safely, economically and efficiently—through the provision of facilities and seamless service in collaboration with all parties and involving airborne and ground-based function." With a simple saying, ATM can be treated as serials of systematic and complex operation on a huge system, which manages countless aircraft and airlines.

ATM encompasses all systems and phases that assist every aircraft to depart from an aerodrome, transit airspace, and land at a destination aerodrome, including air traffic control (ATC), aeronautical meteorology, air navigation systems (aids to navigation), airspace management, air traffic services (ATSs), and air traffic flow management, or air traffic flow and capacity management.

The increasing emphasis of modern ATM is on interoperable and harmonized systems that allow an aircraft to operate with minimum performance change from one airspace to another. ATC systems have traditionally been developed by individual states that concentrated on their own requirements, creating different levels of service and capability around the world. Many air navigation service providers (ANSPs) do not provide an ATC service that matches the capabilities of modern aircraft, so ICAO has developed the Aviation System Block Upgrade initiative in order to harmonize global planning of technology upgrades.

In the past decades, the currently used ATC system faces enormous challenges around the world; the development of ATC lags behind the increasing traffic and forms a bottleneck. The most obvious is a wide range of flight delays that are mainly caused by complex landing paths and waiting procedures; although safety is the primary consideration, the effectiveness of flight is a disadvantage that cannot be ignored. Flight delay will lead to huge economic losses and strong dissatisfaction of travelers. Excessive working pressure of a controller also can cause serious security risks, since the approaching and landing are almost dependent on the organization of the controller and are not automatic; when the traffic flow is in a very high density, mistakes and emergencies are inevitable.

After entering the new century, the civil aviation world has realized the aforementioned problems; also, many projects and researches have been carried on in recent years, such as the following two famous projects:

1. Single European Sky ATM Research (SESAR). It is an ambitious initiative launched by the European Commission in 2004 to reform the architecture of European ATM. Its aim is using modern technology to construct a smart aviation traffic environment in the future.
2. The Next Generation Air Transportation System (NextGEN) is the name given to a new national airspace system due for implementation across the United States in stages between 2012 and 2025.

It can be inferred from SESAR and NextGEN that integration, interoperability, and seamlessness are the most important mechanisms in the further decades to build a modern ATM system. Considering the bulky structure of ATM, in order to achieve these three targets, many new technologies are used, including from physical radio frequency (using higher frequency such as L band to avoid crowed in current VHF band), new wireless data link (e.g., L band digital aeronautical communication system), MIPv6-based mobile network, new surveillance technology (using automatic dependent surveillance [ADS]—broadcast to replace secondary radar), and new computer systems, new management application software, and so on.

The information acquired by various components and departments is integrated by efficiently using a system-wide information management (SWIM) system. The core aim of SWIM is sharing various types of information to support the implementation of collaborative decision making (CDM). There are several important criteria to validate the ATM system, such as real-time data sharing, running from terminal gate to terminal gate, 4D flight path management, flexible and dynamic airspace allocation, and collaboration between space and ground. Figure 1.1 gives a network-based ATM system structure.

As shown in Figure 1.1, there are four subsystems in ATM, that is, sky/ground integrated ATS network, network precision air navigation system, system-wide multilevel monitoring system, and collaborative working ATC system (please note that given the continuous innovation, the name and duty of these subsystems could be changed). The connection between system and people is all based on a network, which is the infrastructure of the ATM system. In the past decades, aeronautical fixed telephone network was mainly used by ATM to exchange various types of message, but there is a trend that the IPv6-based Aeronautical Telecommunication Network (ATN/IPS) will be a promising technology in the future.

For better understanding of Figure 1.1, a real world scenario is given in Figure 1.2.

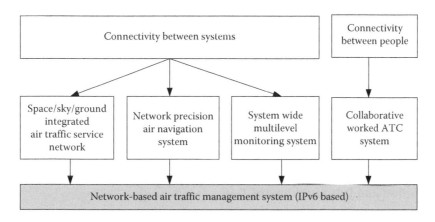

FIGURE 1.1
Framework of network-based air traffic management system.

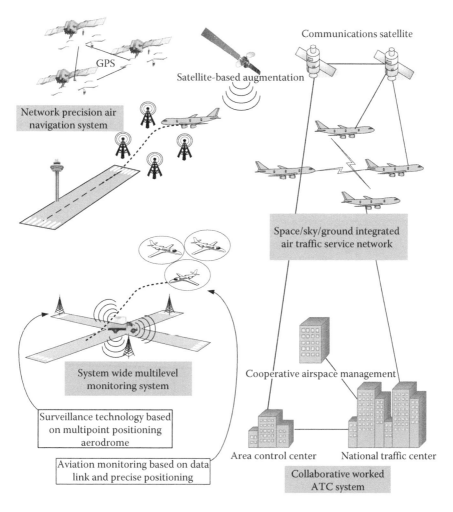

FIGURE 1.2
The typical running scenario of an air traffic management system.

As shown in Figure 1.2, numerous entities participate in the ATM system, such as space relay/GPS satellites, aircrafts, terrestrial data link, and some management entities. All time information exchanging is the critical point that can ensure the running of the whole system. Another point that should be raised naturally is the decision process that is based on the information from the ATM system, which is the CDM in Section 1.2.

1.2 CDM in Air Traffic Management

1.2.1 Definition and Introduction of CDM

The definition of CDM appears in ICAO's document DOC 9971 [4], that is, "CDM is defined as a process focused on how to decide on a course of action articulated between two or more community members. Through this process, ATM community members share information related to that decision and agree on and apply the decision-making approach and principles."

From the top-level viewpoint, the object of this collaborative process is to enhance the holistic performance of the ATM system while balancing the needs of ATM community individual members, for example, airport operators, aircraft operators, ground handlers, and air traffic controllers.

CDM can be applied to all time frames of decisions, from long-range planning of schedules to the tactical decisions of ground delay programs. Each user can participate to a certain level that suits their operations and information requirements. However, in order to maximize the benefits, it is important that all affected users participate in the information sharing, and any member can propose a solution. The members that are commonly included in CDM and their working flow are shown in Figure 1.3.

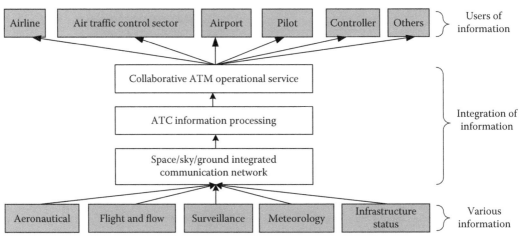

FIGURE 1.3
The collaborative work flow of collaborative decision making.

As shown in Figure 1.3, we can see that CDM is a process of decision based on five types of information: (1) aeronautical information, standards for aeronautical information would be described through the aeronautical information exchange model (AIXM); (2) flight and flow information, the evolution of Doc 9965 [5] would provide initial material to define standards for the flight information exchange model; (3) surveillance information, standards for the ground-to-ground exchange of surveillance information exist today; (4) meteorological information, standards exist for the global dissemination of weather products; further standards development work may be required for new aviation weather products and to make these applicable to aviation CDM (e.g., see WXXM); and (5) infrastructure status, standards for infrastructure status could largely be expressed using modified AIXM standards.

Various data will be utilized by airlines, controllers, aircrafts, and so on. In one word, rational decision should be based on rich and precise information. Some key technologies are used in CDM to ensure this target, such as flexible use of airspace, approach and departure route planning technology for terminal area or multiple airports, sector boundary division technology, capacity assessment and airspace analysis technology, flight procedure design based on multitype navigation, and theory and method for flight procedure assessment.

1.2.2 Common Procedure of CDM

In a collaborative process, the goal is achieving a desired outcome in the most efficient and effective way for the organizations and for all collaborating parties involved. This can only be achieved if the collaborating parties give as much attention to how they work together throughout the process as they do to the process itself. Without one or the other, true cooperation, synergy, and teamwork, the goal can't achieved.

To better understand CDM, some identification should be given: (1) actors ("Who is participating in the collaboration?"); (2) roles and responsibilities ("What functions do the actors perform and how do they interact?"); (3) Information requirements (description of requirements and standards imposed on information exchanged as part of the aforementioned interactions); (4) making the decision ("How is a decision made?"); and (5) rules ("What are some rules constraining the behavior?").

The detailed steps can be summarized as six phases, that is, (1) need identification, (2) analysis, (3) specification and verification, (4) performance case, (5) implementation and validation, (6) operation, and finally subsequent and continuous maintenance and improvement.

The Six Phases of CDM

The first phase is the identification of the need to apply CDM to realize a performance improvement. This can relate to current processes/operations or to future processes. A need statement should refer to the process(es) to which CDM should be applied and need to specify the current situation, involving community members and current (or projected) performance shortfall(s). It should include a first assessment (often based on expert judgment) as to how and through which means CDM can mitigate the shortfall. Shortfalls should be identified in areas related to all 11 key performance areas (KPAs) identified in Doc 9854 [6]. While CDM has the ability to influence performance in all 11 KPAs, CDM provides a mechanism specifically well suited to addressing performance areas, which are frequently difficult to quantify.

In the second phase, the CDM analysis, the target process is further analyzed from a decision-making perspective. The analysis should make clear what decisions to be made, which community members are involved (or affected), which information is used in support of taking the decision(s), which process(es) is followed, how and through which means the decision-making process can be improved, and how such an improvement could contribute to a better performance.

The third phase, which builds on the CDM analysis, results in a shared and verified specification of the CDM process. It will address among others

1. The decisions to be made and how they are reached and finalized
2. The community members involved and their roles/responsibilities in the decision(s)
3. Agreement on objectives, where there may be a shared objective with individual subobjectives (e.g., resolve congestion while minimizing impact to the operation)
4. Decision-making rules, processes, and principles including specification of timeline/milestones, interactions, roles, and responsibilities
5. Information requirements including data standards, quality, frequency, and deadlines
6. CDM maintenance process: review, monitoring/verification, etc.

The objective of the performance case, developed through the fourth phase, is to justify the decision to implement the CDM process and to make the necessary investments. It should clearly specify what the costs are and the benefits (in the relevant KPAs) that will result from the operation of CDM. It is important that the results of the performance case are shared between all relevant community members. In the event that the CDM process is an integral part of a new process, it should be integrated in the performance case for that process.

The fifth phase, CDM validations and implementation, includes all steps to bring CDM into operation. It includes training and informing staff and implementing/adapting systems and information networks.

Once the CDM process is operational (phase six), it should be subject to a continuous and shared review and maintenance and improvement process. In this way, performance can be continually improved.

Source: Adapted from ICAO, *Manual on Collaborative Decision-Making (CDM)*, Doc 9971: AN/485, 2012.

1.3 Aeronautical Telecommunication Network

As aforementioned, it is clear that space/sky/ground communication and information sharing should abide by the principles effectively and safely; ATN that will be introduced in this section is such a framework that can be used to ensure ATM and CDM operations.

ATN is a network that is used by air traffic authorities. It is a global aviation standard telecommunication network established by ICAO, and it is intended to provide seamless air–ground and ground–ground. It uses digital data links to supplement voice communications and provide developed ATSs. The ATN manages the digital data transfer between aircraft and the civil ATC facilities. The current infrastructure of ATN uses the Open Systems Interconnection (OSI) Connectionless Network Protocol.

ATN is a special aviation network that serves ATC units, aviation enterprises, and airline companies. ATN is constituted by the following parts [1]:

1. *Air/Ground Communication Service Provider*: It is responsible for providing the air–ground access networks through terrestrial and satellite data link technologies.

2. *ANSP*: It is responsible for managing the air traffic of one nation or region.

3. *Airline operation*: It manages the commercial operation of an airline company.

4. *Mobile router*: It is responsible for providing access to the air–ground data links for the communication nodes of an airline.

5. *Home agent*: It is a router or routing domain (RD) located at the home link of a mobile node.

6. *ATS/Aeronautical operational control correspondent node*: An ATS correspondent node can be controlled to manage some specific airspace or other nodes to provide information such as weather information. An Aeronautical Operational Control Correspondent Node is located at an airline company or an airport to provide services such as flight arrangement.

In an ATN route approach, the entire ATN is divided into many zones (ATN islands). In each ATN island, there are several backbone RDs responsible for providing the default routes to all the airplanes inside the island. A RD is the first layer default path provider in ATN and is always able to achieve the paths reaching all airlines by some mechanisms. The second default route provider is ATN Home. ATN Home is also a backbone RD that provides the default paths reaching all airplanes of some airline companies. Indeed, ATN Home can get the paths to all airlines of an airline company by some mechanism too [7–9].

ATN consists of a set of end systems (ESs) and a set of intermediate systems (ISs). ESs are the source and destination of all data and are where the applications reside. ISs are better known as routers and relay protocol data units from one system to another. The ESs and ISs are organized into *RDs*. RDs are used to define sets of systems into clusters. These clusters have two major properties: first, they are controlled by a single administration/organization; second, a significant amount of the traffic is internal to the cluster [10].

Today, aeronautical communications use communications radio navigation and surveillance bands and aeronautical radio spectrum as shown in Figures 1.4 and 1.5, respectively [11], while Figure 1.6 shows the hypothetical ATN. The backbone router system is used in designing the ATN to restrict the distribution of aircraft path information in order to avoid the problems of flooding. However, it still allows every ground system the ability to send data to any aircraft. There are four air–ground applications standardized by the international standardization of ATN [12]:

1. *Digital flight information service*: Allows pilots to request continual updates about flight conditions (such as weather)

2. *Controller pilot data link communications* (*CPDLC*): Allows replacing the present controller/pilot voice interactions with digital messages to the extent possible

3. *Context management* (*CM*): Allows supporting a dictionary service that lets the aircraft log in whenever it enters a new ATS authority, exchange applications, and associated network address information with the ground CM application server

4. *ADS*: Allows position information determined by onboard navigation equipment to be transferred to the ground.

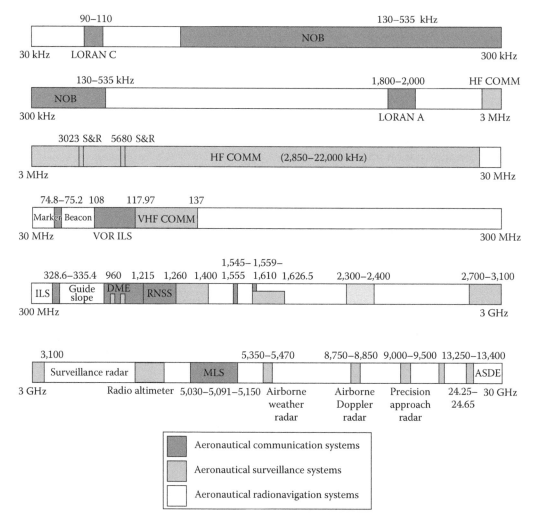

FIGURE 1.4
(See color insert.) Communications radio navigation and surveillance bands.

ATN based on the Quality of Service (QoS) can be divided into four categories [9]:

1. *Residual error probability and transit delay*: The routing decisions should favor low residual error probability over low transit delay.

2. *Residual error probability and cost*: The routing decision should favor low residual error probability over low cost.

3. *Transmit delay and cost*: The routing decisions should favor low transmit time over low cost.

4. *Sequencing and transmit delay*: The routing decisions should favor sending all packets to the destination over a single path to maintain sequencing.

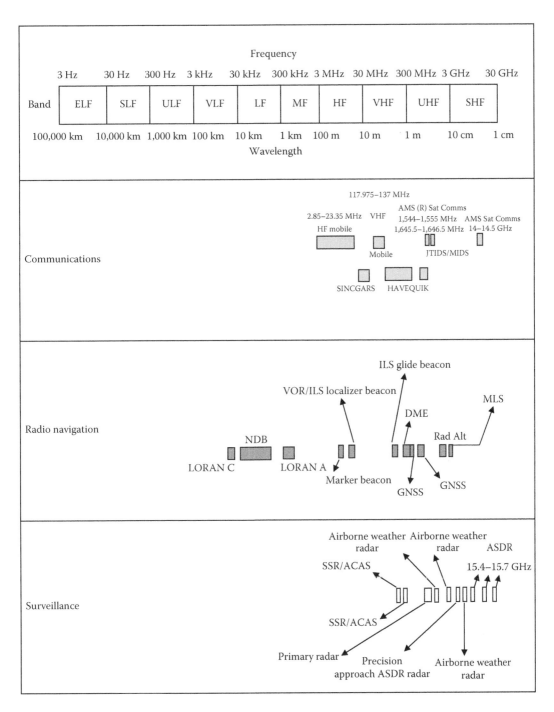

FIGURE 1.5
(See color insert.) Aeronautical radio spectrum.

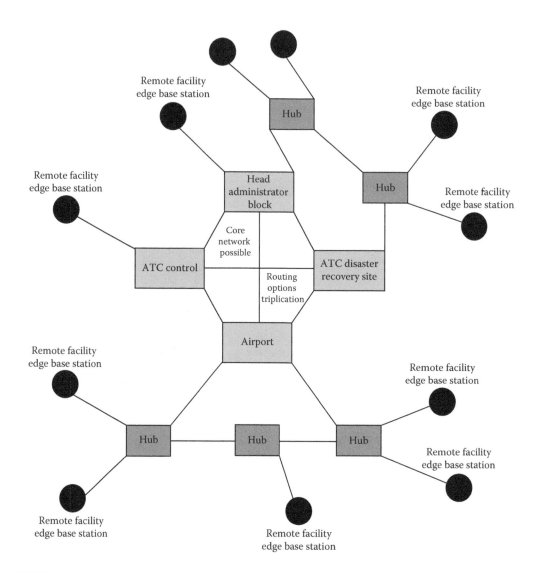

FIGURE 1.6

(See color insert.) Hypothetical *Aeronautical Telecommunication Network.* (From ICAO, *Aeronautical Telecommunication Network (ATN): Manual for the ATN Using IPS Standards and Protocols*, Doc 9896. September 2008.)

ATN is based on the Internet protocol suite (IPS) ATN/IPS [13]. The ATN/IPS protocol architecture is illustrated in Figure 1.7. This architecture has four abstraction layers: the link layer, the Internet layer, the transport layer, and the application layer.

As the ATN/IPS uses the 4-layer model of the Internet Engineering Task Force, Figure 1.8 depicts the relationship between the OSI, ATN/International Organization for Standardization, and the ATN/IPS protocols.

The ATN is enabler for the Communications Navigation and Surveillance (CNS)/ATM. CNS/ATM integrates multiple communication and surveillance system such as GPS and CPDLC.

There are several key technologies that should be further studied in ATN. They are listed in Table 1.1. These technologies are separated by layer for better reading.

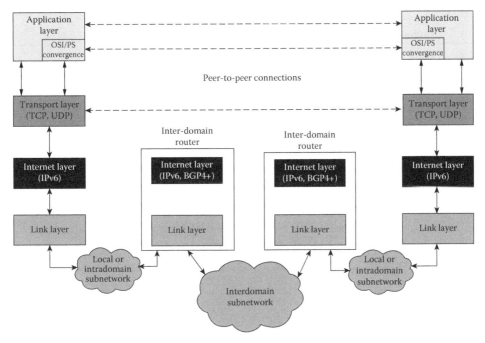

FIGURE 1.7
(See color insert.) Aeronautical Telecommunication Network/Internet protocol suite protocol architecture. (From ICAO, *Aeronautical Telecommunication Network (ATN): Manual for the ATN Using IPS Standards and Protocols*, Doc 9896, September 2008.)

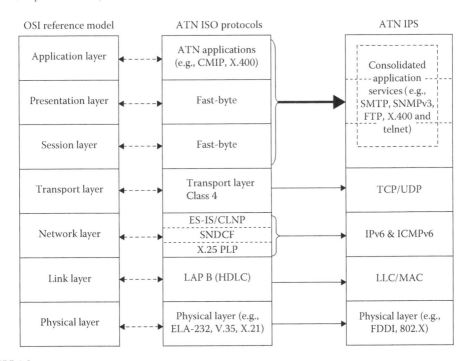

FIGURE 1.8
Protocol reference model. (From ICAO, *Manual for the ATN Using IPS Standards and Protocols*, ICAO, Montreal, Quebec, Canada, 2011.)

TABLE 1.1

Key Technologies in ATN

Name of Layer	Key Technologies
Transport layer	Service-oriented transmission control method in aerospace network environment
	1. Performance analysis of cooperative aerospace information transmission
	2. Qos routing mechanism based on a stable path
	3. Service-oriented aerospace information transmission network control
Network layer	QoS routing technology for high dynamic satellite network topology that includes
	1. Network topology state prediction model
	2. The estimate model for network link status information
	3. Multiconstrained multipath routing algorithm based on the principle of maximum probability
Link layer	High-speed mobile aviation node broadband access technology that includes
	1. Fast speed mobile node access control
	2. Radio resource allocation to meet QoS requirements
	3. Fast switching node based on the minimum cost

References

1. Cook, A. et al. Applying complexity science to air traffic management. *Journal of Air Transport Management* 42, 49–158, 2014.
2. Di Gravio, G. et al. Overall safety performance of air traffic management system: Forecasting and monitoring. *Safety Science* 72, 351–362, 2015.
3. International Civil Aviation Organization (ICAO). *Procedures for Air Navigation Services: Air Traffic Management*, Doc 4444: ATM/501, 15th Edition, 2007.
4. International Civil Aviation Organization (ICAO). *Manual on Collaborative Decision-Making (CDM)*, Doc 9971: AN/485, 2012.
5. International Civil Aviation Organization (ICAO). *Manual on Flight and Flow Information for a Collaborative Environment (FF-ICE)*, Doc 9965: AN/483, 1st Edition, 2012.
6. International Civil Aviation Organization (ICAO), *Global Air Traffic Management Operational Concept*, Doc 9854: AN/458, 1st Edition, 2005.
7. Qingbo, L., Xuejun, X., and Liang, Z. A mobility management mechanism in aeronautical telecommunication network, *IET International Communication Conference on Wireless Mobile and Computing (CCWMC 2009)*, pp. 594–597, 2009.
8. Gallegos, S. K., Link, W. B., and Thomson, D. Innovation approaches for implementing the aeronautical telecommunication network, *12th AIAA/IEEE Digital Avionics Systems Conference (12th DASC)*, pp. 98–103, Fort Worth, TX, 1993.
9. Feighery, P., Hanson, T., Lehman, T., Mondrus, A., Scott, D., Signore, T., Smith, R., and Uhi, G. The aeronautical telecommunications network (ATN) testbed. *15th AIAA/IEEE Digital Avionics Systems Conference*, Atlanta, GA, pp. 117–122, 1996.
10. International Civil Aviation Organization (ICAO), Aeronautical telecommunication network (ATN) architecture plan for the ICAO Africa-India ocean (AFI) region, March 2012.
11. Stacey, D. *Aeronautical Radio Communication Systems and Networks*. John Wiley & Sons, Ltd., Hoboken, NJ, 2008.
12. Signore, T. L. and Girard, M. The aeronautical telecommunication network (ATN). *IEEE Proceedings of Military Communications Conference (MILCOM 98)*, vol. 1, pp. 40–44, Boston, MA, 1998.
13. International Civil Aviation Organization (ICAO). *Aeronautical Telecommunication Network (ATN): Manual for the ATN Using IPS Standards and Protocols, Part I, II, III*. ICAO, Montreal, Quebec, Canada, 2011.

2

Optimization and Enhancement of MIPv6 in ATN

Douzhe Li and Zhao Li

CONTENTS

2.1 Introduction

The next-generation Aeronautical Telecommunication Network (ATN)[1] is an all-IP (Internet protocol) mobile communication network (abbreviated as ATN/IPS, here IPS means Internal protocol suite). It will also conform to the concept of e-enabled aircraft in the next 30 years.[2] In the network layer, Mobile IPv6 (MIPv6) is selected as its mobile communication/management protocol. MIPv6 allows mobile nodes (MNs, here refers to aircraft stations) to remain reachable while flying around in the IPv6 network. Each MN is always identified by its home address (HoA), regardless of its current point of attachment to the network. While located away from its home, an MN will be associated with a care-of address (CoA), which is configured by the attached network and provides information about the current location of MNs.

In this chapter, first, some major concepts of general-purpose MIPv6 are described in Sections 2.2 and 2.3. Then, we will combine the MIPv6 with its specific application in ATN, especially elaborate some problems or enhancements of MIPv6 that should be focused on.

One major problem of MIPv6 is triangular routing, which depends on the relative location of the home agent (HA); there can be a situation where the path in one direction (e.g., correspondent node [CN] to HA to MN) is significantly longer than the path in the reverse direction (e.g., MN to CN). Detailed explanation and some solution schemes and relative analytical model will be elaborated in Section 2.4.

Enhancements of MIPv6 such as FMIPv6, HMIPv6, and PMIPv6, which are used to improve handover latency, reduce the amount of signaling required, and facilitate different access technology, will be introduced and elaborated in Section 2.5, respectively.

Network mobility (NEMO) is also treated as an extension of MIPv6 and will allow any node in the network that moves as a unit. It is very useful when MIPv6 is deployed in a nautical environment. But an apparent drawback of NEMO is that it does not support route optimization (RO), a feature that provides better end-to-end delay and path length (in terms of number of hops); the optimization methods and related analysis will be given in Section 2.6.

2.2 Development of ATN Wireless Communication Standard

There are two organizations that formulate mainly relevant standards, that is, the European Organization for the safety of air navigation (EUROCONTROL) and the International Civil Aviation Organization (ICAO).

EUROCONTROL's Statistics and Forecast Service (STATFOR) has the goal of providing statistics and forecasting on air traffic and monitoring the evolution of the air transport industry. STATFOR provides short-term, medium-term, and long-term forecasts. Thereby, different growth scenarios have been developed in order to account for different economic progresses. The latest long-term forecast that was published in 2013[3] predicts air traffic growth up to the year of 2050; according to the regulated growth scenario (i.e., growth scenario C), it represents an extension to the existing environment we are in today. It is characterized by moderate economic growth, with regulation reconciling the environmental, social, and economic demands to address the growing global sustainability concerns. It exhibits a medium level of growth with 18.6 million Instrument Flight Rule movements in Europe by 2050—2.0 times more than in 2012.

ICAO has recognized the need for technological progress. Specifically, this has been discussed at the 11th Air Navigation Conference, where the conclusion was that the aeronautical mobile communication infrastructure had to evolve in order to accommodate new functions of air traffic management (ATM) and to support air traffic growth. As a result, EUROCONTROL and the U.S. Federal Aviation Administration decided to establish a dedicated working arrangement (i.e., Action Plan 17,[4] AP17). The objective of that working group was the first step toward an enhanced aeronautical communication infrastructure, generically called Future Communication Infrastructure (FCI). The FCI includes all components that are necessary for Air Navigation Service Provider

(ANSP), aeronautical operational centers, and aircraft to communicate with each other. Thereby the so-called Future Radio System (FRS) is a subpart of the FCI. The FRS is a term introduced by the Communications Operating Concept and Requirements for the Future Radio System (COCR) document[5] and indicates a globally consistent technological solution that provides appropriate communication infrastructure in order to accommodate anticipated future air traffic growth. Thereby, the FRS is not necessarily a single technology. AP17 has been a joint activity between Europe and the United States, where several European and American partners worked in close cooperation in order to assess and harmonize possibilities and solutions that would be appropriate for the modernization of the aeronautical communication infrastructure. Thereby, it was ascertained that the future ATM will face a paradigm shift from voice-based toward data-based operations. AP17 was divided into several tasks, where six of them were technically related. In the context of this chapter, the third task, *investigation of new technologies for mobile communication* is of special interest. This technical task investigated newly emerging and existing data link technologies suitable to fulfill the requirements of the FCI. Within this task, several technologies for different operational aspects were compared.

A concurrent activity conducted by Working Group I of ICAO's Aeronautical Communication Panel (abbreviated as ACP-WG-I) was the development of the ATN/IPS protocol suite. Members of AP17 also concluded that the FCI will employ an ATN/IPS infrastructure. After the selection of the new communication systems has been made (which is still not determined at the time of this book), the required Quality of Service (QoS) shall be reassessed within an end-to-end IP environment. Since commercial-off-the-shelf (COTS) products can facilitate development and ensure robustness, AP17 also recognized that it would be beneficial if COTS products could be used or parts of COTS could be reused for future aeronautical communication systems. It is crucial to realize the importance of that statement. It means that COTS products shall be used wherever possible; however, if not applicable, components of COTS products shall be reused if workable.

Some key recommendations were given in AP17: the development of an airport surface communication system (i.e., AeroMACS),·to finalize the selection of a terrestrial data link communication system (i.e., L-DACS)[6], and to consider satellite-based communication systems, the prime candidate for oceanic and remote areas (c.f. ESA Iris). Furthermore, AP17 concluded that it is important to retain QoS mechanisms provided by any newly introduced data link technology when integrating the data link into a subnetwork. This is imperative when considering the operation of IP over a specific aeronautical link.

2.2.1 Framework of Aeronautical Mobile Communication Network

These data transmissions formed a network that contains wire and wireless communication. Cellular network has been already used in various wireless communication systems, for example, personal mobile phone or police radio cluster communication. The cellular concept can be applied to an aeronautical scenario in the same way. During the process of a commercial flight, there are about five stages, namely, taxing, parking, takeoff, cruising, and landing. Transmitting data, voice, and other useful information between aircraft and ground stations is almost dependent on wireless communication. So an aeronautical mobile communication network, allowing transmission of bidirectional data via a terminal device, need not be connected to a fixed physical link.

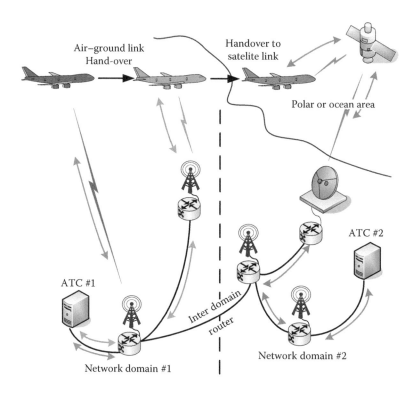

FIGURE 2.1
(See color insert.) ATN mobile communication framework.

Controller–pilot voice communication is widely established throughout the aeronautical world and has had a very rapid increase in the number of subscribers to the various ground station cellular networks over the last few years, although most of them are still using analog HF band.

Aeronautical data communication has become a very important and rapidly evolving technology as it allows users to transmit data from remote locations to other remote or fixed locations. This proves to be the solution to the biggest problem of monitoring aircraft status and supplying an Internet connection to traveling business people. Figure 2.1 gives an illustration of the overall environment of ATN.

As we can see from Figure 2.1, different colors (orange/blue/green) indicate different locations of an airplane. When the aircraft is flying over a continental area, the aircraft station communicates with air traffic control (ATC) center #1 by using air-to-ground data link, along with the path: airborne router → ground station → ground router (the orange and blue routing path). Communication between planes can utilize air-to-air data link (air-to-air link is out of scope of this chapter). In remote areas such as ocean and polar, you can access ATN network via airborne router → satellite station → satellite data link → satellite ground station (green routing path). In addition, each ground station can communicate only with limited distance and a limited number of aircrafts, so handover between different ground stations is necessary.

The aforementioned is a perceptual introduction. Here we give a further presentation of the communication protocol suite of aeronautical mobile communication network shown in Figure 2.2.

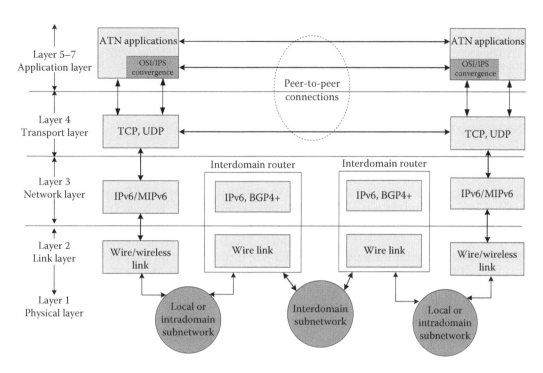

FIGURE 2.2
Protocol architecture of ATN/IPS.

In Figure 2.2, the ground station, aircraft, satellites, or ATC centers (ATC#1, ATC#2), all can be treated as IPS nodes, which have different responsibilities and functions. The access and interconnection of ATN are realized by an IPS node, which is a device that implements IPv6/MIPv6. There are two types of IPS nodes: (1) an IPS router, which forwards IP packets not explicitly addressed to itself, and (2) an IPS host, which can be treated as a message source or a message destination, and it does not forward any packets.

We will now describe the internal structure of ATN/IPS from three viewpoints.

From the viewpoint of physical topology, the ATN/IPS internetwork is composed of IPS nodes and a huge network which operate in a worldwide environment. The main function of ATN is supporting air traffic service communication (ATSC) and aeronautical industry service communication which contains aeronautical administrative communications (AAC) and aeronautical operational communications (AOC).

From an administrative viewpoint, the ATN/IPS internetwork is composed of numerous interconnected administrative domains. The concept of administrative domain is abstract, but realistically, it may be an individual state, or a set of states that form an ICAO region, an Air Communications Service Provider (ACSP), an ANSP, or any other organizational entity that has the authority to manage ATN/IPS network resources and services. In each administrative domain, there must exist several (also may be only one) IPS routers which execute the interdomain routing protocol (such as the famous Border Gateway Protocol version 4) specified in the ATN/IPS manual. Also, administrative domains and their counterparts should coordinate their policy for carrying transit traffic.

From the viewpoint of routing, the routing information between autonomous systems (ASs) can be exchanged by interdomain routing protocols. An AS is a connected group of one or more IP address prefixes. The routing information exchanged includes IP address prefixes of different lengths. For instance, the length of the IP address prefix exchanged between ICAO regions may be shorter than the IP address prefix exchanged between individual states within a particular region.

Mobility service providers (MSPs) will operate ATN/IPS mobility, which is based on IPv6 mobility standards. An MSP in the ATN/IPS is an instance of an administrative domain, which may be an ACSP, ANSP, airline, airport authority, government, or some aviation organization. ATN/IPS MSP shall operate one or more HAs.

2.2.2 MIPv6-Based Mobile Communication

As we already know the mobility is braced by the IPS (i.e., IPv6 and MIPv6) protocol suite, first, we give an intuitive metaphor. For better communication between nodes, each node like the mailbox in front of our house should be identified with a mailing address and the ZIP code. During mail delivery, the postman checks the address and then delivers mail to the mailbox. Detailed information about IPS will be discussed as follows.

As proposed by the Internet Engineering Task Force (IETF) in RFC 2002 and subsequent RFCs, mobile IP can provide an efficient, scalable mechanism for node mobility within the Internet. Nodes may move and change their points of attachment to the Internet without changing their IP address. This allows them to remain transparent to higher-layer connections while moving. Node mobility is realized without the need to propagate host-specific routes throughout the Internet routing fabric. The MN uses two IP addresses: a fixed HoA and a CoA that changes at each new point of attachment. This mechanism is briefly illustrated in Figure 2.3.

The current most implemented version of mobile IP is version 4 (MIPv4), which assumes that a node's IP address uniquely identifies the node's point of attachment to the Internet. A node must be located on the network indicated by its IP address in order to receive datagrams destined for it; otherwise, datagrams destined to the node would be undeliverable. If a node changes its point of attachment, in order not to lose its ability to communicate, one of the two following mechanisms must typically be employed:

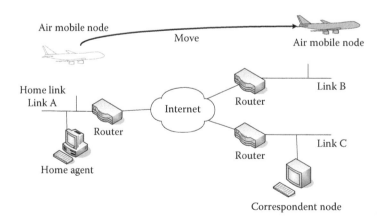

FIGURE 2.3
Illustration of mobility and transparent data transmission.

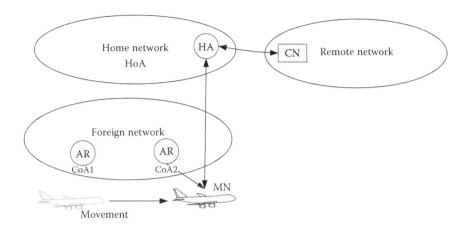

FIGURE 2.4
Simplified mobile IP.

1. The node must change its IP address whenever it changes its point of attachment.
2. Host-specific routes must be propagated throughout much of the Internet.

The first alternative makes it impossible for a node to maintain transport and higher-layer connections when the node changes location. The second does not scale very well.

According to mobile IP protocol, an MN has two addresses: an HoA, which is a permanent address, and a dynamic CoA, which changes as the MN changes its point of attachment (See Figure 2.4; it is a simplified version of Figure 2.3). The fundamental technique of mobile IP is forwarding. A CN, which is any peer node with which an MN is communicating, sends packets to the HA of the MN. The CN reaches the HA through normal IP routing. Upon receipt of a packet from the CN, the HA will forward these packets to the MN at its current CoA. The HA simply tunnels the original packet in another packet with its own source address and a destination address of the current CoA. This is possible because of the mobile IP protocol whereby the MN sends *binding update* (BU) messages to the HA whenever its point of attachment changes. The BU associates the MN's HoA with its current CoA. The HA maintains a *binding cache* that stores the current CoA of the MN.

In order to describe the different functional building blocks of a typical aeronautical wireless communication system, a generic functional architecture is used. Such a description consists of several entities that implement different functions. Thereby, entities may be physically separated but may also be hosted by a single system. The various entities communicate among each other via interfaces (also known as reference points). These interfaces are used to exchange user data (i.e., via the bearer plane) or signaling data (i.e., via the control plane). The relevant part for the operation of IP over an aeronautical link technology is how the bearer plane from the mobile router (MR) to the access router (AR) is realized.

Figure 2.2 shows a generic functional architecture of a wireless communication system with centralized medium access. In principle, each aeronautical communication system could be mapped to such a generic functional architecture. The mobile station, which is hosting an IPv6-capable router, is connected to the ground infrastructure of the communication system via the wireless link to the base station (BS). Within this context, the entity that is controlling medium access to the physical transmission medium is generically called the base station (BS). The wireless link itself is specifically designed for its purpose and is specified through wireless interface specifications. Thereby, a link comprises the physical layer and data link layer (DLL) definition. Mostly, such a standard includes a rudimentary definition of an interface toward upper layers (i.e., the Internet layer), too.

A BS may communicate with other BSs through a specific interface. An inter-BS interface may be interesting in order to speed up BS handovers, but it may also be interesting for a DLL network. The BSs themselves are in general connected to a link-specific access gateway (LSAG) that is responsible for a domain as depicted in Figure 2.2. A domain within this context is understood as a subnetwork of a larger network (i.e., the ATN). Thereby, the LSAG may host other entities that are responsible for signaling, handovers, or resource management. Furthermore, an AR is either directly attached to the LSAG or reachable via an additional interface.

Theoretically, there may be a single BS responsible for a whole domain, especially if satellite-based communication is considered. Ideally, a link technology design specifies a service that fits exactly the needs of the upper layers. That is, the Internet layer uses the specific link technology transparently, and all protocols implementing the service specification of the Internet layer work properly. However, because network access providers, content providers, and customers have very different requirements for networks, it is hardly possible to satisfy all needs accordingly. It might appear that assembling all requirements for a safety-related aviation case would be simpler than in other commercially driven use cases, as requirements should be clearly defined and available. However, due to the fact that the aviation industry is evolving rather slowly (at least from a technological point of view), it is hard to nail down all requirements for projects probably realized 20 years in the future (if at all). Nevertheless, from the network point of view, one hard requirement that should definitely be met by all future communication systems is full IPv6 compliance.

By strictly following this requirement, any future amendment of the ATN/IPS shall be of no concern. Furthermore, it would be beneficial if IP multicast could be supported in an efficient manner.

2.3 Basics of Mobile IPv6

2.3.1 MIPv6 Protocol

Before describing the working mechanism of mobile IPv6 (MIPv6) and its improvement, some important terms in MIPv6 are adapted[7] as follows.

- *Mobile node* (MN): A IPv6 node that is identified by its home address (HoA) disregarding its current location. However, the MN should maintain reachability using its HoA. When moving in the network, the MN may change its access links, and then it will be associated with a new care-of address (CoA). By utilizing the CoA, an MN has the awareness of its global address for the link that it is attached. HoA and CoA is binded, and a tunnel is constructed between the HA and MN for communication.

- *Foreign link*: A link that is in the MN's visiting network, and is not the MN's home link.

- *Care-of address* (CoA): CoA is associated with the network that the mobile node is visiting and provides information about the mobile node's current location. The entire datagram addressed to the MN's HoA will be transparently forwarded to the CoA. The CoA includes the foreign subnet prefix and an interface ID determined by the MN. This mechanism can realize the stateless address configuration. MIPv6 allows the MN to be associated with multiple CoAs, but at the same time, there is only one "primary" CoA, which is the one registered with the MN's HA.

- *Correspondent node* (CN): An IPv6 node that communicates with an MN. A CN may not have a mobile property, and may be a stationary PC. If the CN is an MN, it should have the MIPv6 protocol stack installed and be away from home.
- *Home link*: The link that is affiliated to home subnet prefix. The HoA is within the MN's home link, also the HA always resides on the home link.
- *Home address* (HoA): The permanent address of the MN whenever the MN is in the home network or in a foreign network; thus, HoA is always reachable. It can be understood that the MN is always logically connected to the home link. An MN can have multiple HoAs if there are several home subnet prefixes. When the MN is attached to the home link, it will be operated like an IPv6 host, but when the MN is away from home, the datagram destined to the MN should be forwarded from the HA to the MN by a tunnel.
- *Home agent* (HA): A router resides on the MN's home network. The HA records the MN's current location information (i.e, the CoA). When the MN is away from home, it registers its current address with the HA, which forwards data sent to the MN's HoA to the MN's current address on an IPv6 network through tunnels, and also forwards tunneled data sent by the MN. A home agent may work in conjunction with a foreign agent, which is a router on the visited network.

Although the HA is a router connecting the home link to an IPv6 network, the HA does not have to serve this function. The HA can also be a node on the home link that does not perform any forwarding when the MN is at home.

MIPv6 allows an IPv6 node to be mobile to arbitrarily change its location on an IPv6 network and still maintain existing connections. When an IPv6 node changes its location, it might also change its link. When an IPv6 node changes its link, its IPv6 address might also change in order to maintain connectivity. There are mechanisms to allow for the change in addresses when moving to a different link, such as stateful and stateless address auto-configuration for IPv6. However, when the address changes, the existing connections of the MN that is using the address assigned from the previously connected link cannot be maintained and are ungracefully terminated.

The key benefit of MIPv6 is that even though the MN changes locations and addresses, the existing connections through which the MN is communicating are maintained. To accomplish this, connections to MNs are made with a specific address that is always assigned to the MN and through which the MN is always reachable. MIPv6 provides transport layer connection survivability when a node moves from one link to another by performing address maintenance for MNs at the Internet layer.

2.3.2 Understanding Details of MIPv6

In this section, we use the knowledge gained from previous sections to draw a simplified flowchart for the MN's and CN's operation. This flowchart can be compared to the one used in Section 2.2 for normal IPv6 hosts. Since we want to focus the mobile signaling mechanism, the BU security details were not considered in this section. Our flowchart for the MN's sending implementation is shown in Figure 2.5.

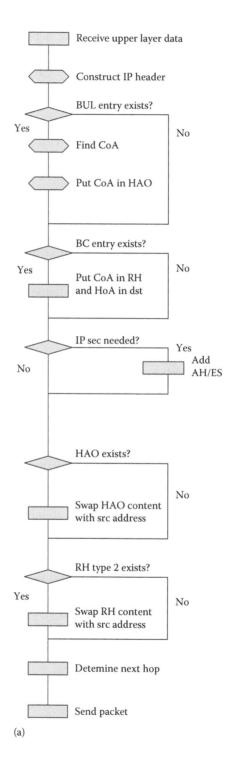

FIGURE 2.5
(a) Processing sending packets at the correspondent node.   *(Continued)*

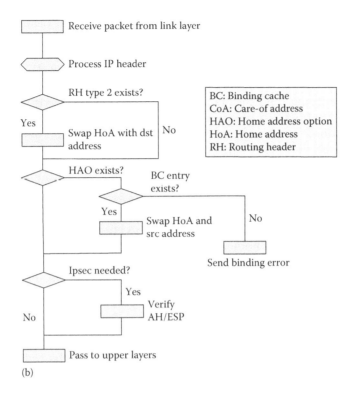

FIGURE 2.5 (*Continued*)
(b) Processing receiving packets at the correspondent node.

When the MN's IP layer implementation receives upper-layer data, it will have already supplied the HoA as a source address for applications. Hence, an IP header can be constructed with the HoA as the source address and the CN's address as the destination address. Next, the BU list is checked to see if the MN has already sent a BU to the CN and get the CoA from the correspondent's node's entry. If it has, the MN can include its HoA in the HoA option. However, for now, the MN places its CoA in the HoA option instead. We see why this is done when we analyze the following steps in the flowchart. The MN then checks its binding cache to see if the CN has sent it a BU. Note that CN refers to a functional entity in an IPv6 node and does not exclude a CN from being an MN as well. If the CN had sent a BU to the MN, a binding cache entry is found. Hence, the MN constructs a routing header type 2 and places the CN's CoA inside it.

Assuming that the MN does not need to add any other extension headers, the next step is to check whether IPsec (IP Security, a protocol suite for securing IP communications, details not covered in this chapter) is needed. This is done by checking the security policy database to see whether the packet should be protected. When IPsec is used, the destination options header containing the HoA option needs to be placed before the IPsec header (AH or ESP). This is done because IPsec identifies a security association in the security association database by selectors. Selectors include the source and destination addresses of the nodes sharing the security association. To avoid changing the security association every time the MN moves, it is best to use the HoA (not the CoA) as a selector identifying the MN. When the packet is received by the CN, the headers are processed in the order they appear in the packet; therefore, IPsec headers are processed last, before the upper layer. Hence, an IPsec implementation in the CN always sees the HoA in the source address field of a received packet. Now we can see the reason for keeping the MN's HoA in the source address field and its CoA in the HoA option until IPsec has added its header. We need to perform the IPsec operations on the header seen by the CN after verifying the IPsec header. That is, the packet needs to look exactly the same when passed to the CN's IPsec implementation as it did when passed to the MN's IPsec implementation.

Otherwise, to IPsec, the packet will seem to have been modified and will be dropped.

After IPsec processing is done, the MN swaps the contents of the HoA option and the routing header with the source and destination addresses respectively. The next hop (the next destination of the data packet) is determined based on whether the packet is tunneled to the HA or sent directly to the CN, then the addresses of next hop are tied to interfaces. For example, when the MN is on a foreign link, its HoA is associated with a tunnel interface to the HA. When it moves back home, the tunnel is deleted, and the HoA is associated with a physical interface. The tunnel interface can be treated by the IP layer implementation as another physical interface that is one hop away from the HA. Hence, if the packet contains the HoA in the source address field, it is immediately tunneled to the HA; otherwise, it is sent on the interface associated with the CoA.

Now let's consider the steps taken by the CN when it receives packets from the MN, as shown in Figure 2.5b.

When the CN receives a packet, it processes the headers in the order that they were sent. If a routing header type 2 is included, and provided that there is one address in it and that it is the CN's HoA, it swaps that address with the destination address in the packet; otherwise, the packet is dropped. The node then processes other extension headers. In Figure 2.5, we skip this processing until it gets to the destination options header. If an HoA option is found, the binding cache is searched for a corresponding entry. If nothing is found, a binding error message is sent to the source address in the IPv6 header. If the CN does not implement any MIPv6 messages, then it does not understand the HoA option and sends an ICMP error message to the source address in the packet. If a binding cache entry is found for this HoA, the home address is swapped with the CoA in the source address field.

If an IPsec header is included, it is verified as usual. Note that the packet would now look exactly the same when seen by IPsec implementations at both ends. Finally, the information is passed to upper layers.

2.3.3 Open-Source Simulation Tools

Open-source network simulator tools are often used in academic researching; although open-source software does not have the privilege of customer service, it's easy for modification, realizing, and testing a researcher's novel idea; also it can help the researcher to understand the whole network framework. Usually there are two famous open-source network simulators mentioned in academic papers, that is, OMNET++ and NS (NS-2 or NS-3).

The NS-3 simulator is a discrete-event network simulator tool, which is typically used for research and educational purpose. The NS-3 project is an open-source project, which was started in 2006, and the current version is NS-3.21 (until Sep. 2014). Some tips as follows can help us understand NS-3.

- NS-3 is not an extension of NS-2. It is a new simulator. Both the simulators are written in C++, but NS-3 is a new simulator that does not support the NS-2 APIs. Some models from NS-2 have already been ported from NS-2 to NS-3. The project will continue to maintain NS-2 while NS-3 is being built, and will study transition and integration mechanisms.
- NS-3 is open source, and the project strives to maintain an open environment for researchers to contribute and share their software.
- In NS-3, the simulation code should be written in C++ (recommended) or Python; NS-3 dropped the TCL simulation code, which was widely used in NS-2. So proficient C|+ programming skills can help to quickly grasp NS-3.

For the purpose of simulating ATN/IPS, there is nothing difficult in downloading and compiling NS-3 for preparation. The importance is currently there is not an official MIPv6 simulation environment in the latest release. MIPv6 and NEMO simulations are base on a third-party component called Usagi-Patched MIPv6 stack (UMIP), which was written by a Japanese programmer.

2.3.3.1 Using Direct Code Execution (DCE)

The DCE NS-3 module provides facilities to execute existing implementations of userspace and kernel space network protocols within NS-3.

The motivation and advantages of utilizing DCE are

- If you do not want to maintain multiple implementations of a single protocol or re-implement complicated protocols such as Open Shortest Path First (OPSF) or Border Gateway Protocol (BGP) routing
- If you want to debug and test your code within a controlled environment
- If you want to create a miniature network (let's say, a model of an ISP network) in a single node

As of today, the Quagga routing protocol implementation, the CCNx CCN implementation, and recent versions of the Linux kernel network stack are known to run within DCE, hence allowing network protocol experimenters and researchers to use the unmodified implementation of their protocols for real-world deployments and simulations.

Using UMIP: UMIP is an open-source implementation of MIPv6 and NEMO Basic Support for Linux. It is released under the GPLv2 license. (Please visit http://umip.org/ for more recent news and updates; latest update is UMIP 1.0.) The UMIP support on DCE enables the users to reuse routing protocol implementations of UMIP. UMIP now supports MIPv6 (RFC3775), NEMO (RFC3963), Proxy Mobile IPv6 (PMIPv6; RFC5213), RFC3776 and RFC4877 (IPsec and Internet Key Exchange version 2nd [IKEv2]), etc., and these protocol implementations can be used as models of network simulation. It reduces the time of reimplementation of the model and potentially improves the result of the simulation since it is already *actively running* in the real world.

Unlike the Quagga support of DCE, the UMIP support of DCE requires Linux kernel integration of DCE, because the dependencies of UMIP, UMIP must interact with Linux kernel.

2.4 Optimization of Triangular Routing in MIPv6

2.4.1 Brief Introduction of Triangular Routing

While running mobile IP protocol, an MN owns two addresses: an *HoA*, which is a permanent address, and a dynamic *CoA*, which changes as the MN changes its point of attachment. The fundamental technique of mobile IP is forwarding. A CN, which is any peer node with which an MN is communicating, sends packets to the HA of the MN. The CN reaches the HA through normal IP routing. Upon receipt of a packet from the CN, the HA will forward these packets to the MN at its current CoA. The HA simply encapsulates the original packet in another packet with its own source address and a destination address of the current CoA. This scheme is realized by BU messages, which are sent by MN to the HA whenever MN's point of attachment changes. The BU associates the MN's HoA with its current CoA. The HA maintains a BC that stores the current CoA of the MN. In the reverse direction, the MN could simply send packets directly to the CN using normal IP routing.

While all traffic from the CN to the MN is routed via the HA, then this situation is called triangular routing. This type of routing increases the traffic on the network as the packets are first routed to the HA, and from here, they are transmitted to the MN by a tunnel. In particular, the load on the HA will be increased significantly. As we can see from Figure 2.6, the triangular routing problem is demonstrated by a very far distance between the MN (i.e., aircraft node) in America and the HA in Asia; all the data from MN to CN should be forwarded by HA. This process will consume much network resource. While with optimization, the data can be directly forwarded to CN.

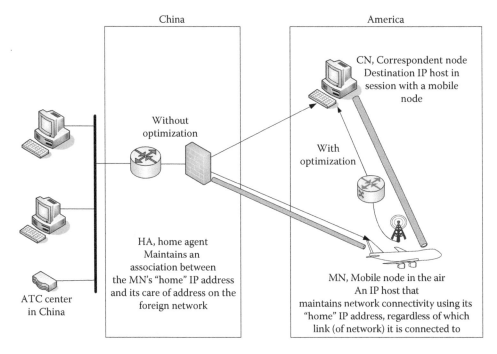

FIGURE 2.6
Solution of triangular routing.

However, this results in *triangular routing*, and depending on the relative location of the HA, there can be a situation where the path in one direction (e.g., CN to HA to MN) is significantly longer than the path in the reverse direction (e.g., MN to CA). A further consideration in this case occurs if the MN uses its HoA as the source address. The problem here is that many networks perform *ingress filtering* of incoming packets and will not accept packets that are topologically incorrect. This would be the case with packets from the MN because they actually originate from the CoA, but the source address in the IP packet is the HoA.

2.4.2 Optimization of Triangular Routing

In the reverse direction, the MN could simply send packets directly to the CN using normal IP routing. However, this results in *triangular routing*, and depending on the relative location of the HA, there can be a situation where the path in one direction (e.g., CN to HA to MN) is significantly longer than the path in the reverse direction (e.g., MN to CN). A further consideration in this case occurs if the MN uses its HoA as the source address. The problem here is that many networks perform *ingress filtering* of incoming packets and will not accept packets that are topologically incorrect. This would be the case with packets from the MN because they actually originate from the CoA, but the source address in the IP packet is the HoA. Because of these issues, MIPv6 allows the MN to follow the same path back to the CN via the HA using *bidirectional tunneling* whereby in addition to the HA tunneling packets to the MN, the MN *reverse-tunnels* packets to the HA. The HA will decapsulate a tunneled IP packet and forward it to the CN. With bidirectional tunneling, the CN is not required to support mobile IP.

Bidirectional tunneling solves the problems of triangular routing and ingress filtering; however, there still can be suboptimal routing since the path from the MN to the CN via the HA may be relatively long even when the MN and the CN are in close proximity. With MIPv6, the situation where the path through the HA is longer than a direct path to the CN may be addressed using a technique called route optimization. With RO, the MN sends BUs to both the HA and the CN. In this case, the MN and the CN can communicate directly and adapt to changes in the MN's CoA. RFC 3775 defines the procedures for RO. It requires that the MN initiate the return routability procedure. This procedure provides the CN with some reasonable assurance that the MN is addressable at its claimed CoA and its HoA.

It is generally acknowledged that there are drawbacks to RO. RFC 4651 presents a taxonomy and analysis of enhancements to MIPv6 RO. This document notes that the two reachability tests of the return routability procedure can lead to a handoff delay unacceptable for many real-time or interactive applications, that the security and the return routability procedure guarantees might not be sufficient for security-sensitive applications, and that periodically refreshing a registration at a CN implies a hidden signaling overhead. Because of the overhead and delay associated with the return routability procedure and because at least for ATSC it is expected that the CN and the HN will be in relative close proximity, this manual requires that IPS CNs that implement MIPv6 RO allow RO to be administratively enabled or disabled with the default being disabled. New solutions to RO are expected as a result of IETF-chartered work in the mobility extensions for IPv6 working group, which includes aviation-specific requirements.

2.5 Enhancement of MIPv6

2.5.1 HMIPv6

One technology for these issues is *hierarchical MIPv6 (HMIPv6)* (RFC 5380). RFC 5380 introduces extensions to MIPv6 and IPv6 neighbor discovery to allow for local mobility handling. HMIPv6 reduces the amount of signaling between an MN, its CNs, and its HA. HMIPv6 introduces the concept: the mobility anchor point (MAP). A MAP is essentially a local HA for an MN.

When an MN visits a new access network and enters a MAP domain, it will receive router advertisements that include information about one or more local MAPs. The MN can bind its current location, that is, its on-link care-of address (LCoA), with an address on the MAP's subnet, called a regional care-of address (RCoA). Actually LCoA is a CoA used in the binding update to MAP, while RCoA is used for route optimization, and it is a CoA in the binding update to home agent and corresponding nodes. MAP can be treated as a local HA, and it will receive all packets in order to the MN it is serving and will encapsulate and forward them directly to the MN's current address. If the MN does not change its current address within a local MAP domain (LCoA), it needs not to register only the new address with the MAP. The RCoA will change only if the MN moves within a MAP domain.

RFC 4140 notes that the use of the MAP does not assume that a permanent HA is present, that is, an MN need not have a permanent HoA or HA in order to be HMIPv6-aware or use the features of HMIPv6. HMIPv6-aware MNs can use their RCoA as the source address without using an HoA option. In this way, the RCoA can be used as an identifier address for upper layers. Using this feature, the MN will be seen by the CN as a fixed node while moving within a MAP domain. This usage of the RCoA does not have the cost of MIPv6 (i.e., no bindings or HoA options are sent back to the HA), but still provides local mobility management to the MNs with near-optimal routing although such use of RCoA does not provide global mobility.

Here we will bring in the concept of a hierarchical MIPv6 network. A new node is introduced named as the mobility anchor point (MAP).

An MIPv6 node can move within the Internet topology while maintaining reachability and ongoing sessions between the MN and CN according to MIPv6 (RFC3775). To fulfill this, every time an MN moves to another point, it should send BUs to its HA.

The time delay often happens in the process of signaling exchange. For example, after sending the BU, the MN may send datagrams by means of its HA immediately, but the HA cannot send any datagrams back to the MN before it receives the BU. This will cause at least half a round-trip delay before packets reach the right place again. We will have another delay event for sending data packets if the MN chooses to wait for a binding acknowledgment (BA) returned. The worst situation always occurs when the MN and its HA are in different and distantly located across the world, so the round-trip times will be more longer.

Additional delay will also occur if the MN adopts a route optimization mechanism. Authenticating BUs requires about 1.5 round-trip times between the MN and each CN. (Assume in the ideal environment, the entire return routability procedure has no packet loss.) We can do it in parallel by sending BUs to the HA, and there will be further optimizations that reduce the required 1.5 round-trips (RFC4449, RFC4651, and RFC4866).

Nevertheless, the signaling exchanges need to update the location and they will always cause some disruption to active connections. Some packets will be lost. There may be effects to upper-layer protocols along with link layer and IP layer connection setup delays. We will enhance the performance of MIPv6 if we reduce these delays during the time-critical handover period.

Moreover, as for wireless links, such a solution reduces the number of messages which are sent over the air interface to all CNs and the HA. A local anchor point will also allow MIPv6 to benefit from reduced mobility signaling with external networks.

For these reasons, a new MIPv6 node with the name of the mobility anchor point is used and can be located at any level in a hierarchical network of routers, including the AR. The MAP will limit the amount of MIPv6 signaling outside the local domain.

The introduction of the MAP provides a solution to the issues outlined earlier, in the following way:

- The MN sends BUs to the local MAP instead of the HA (which is typically further away) and CNs.
- We should transmit the only BU message by the MN before traffic from the HA and all CNs is rerouted to its new location. This is independent of the number of CNs with which the MN is communicating.

Substantially, A MAP is a local HA. If we want to improve the performance of MIPv6 while minimizing the impact on MIPv6 or other IPv6 protocols we should introduce the hierarchical mobility management model in MIPv6. For some security considerations, while using MIPv6 RO, HMIPv6 allows MNs to hide their location from CNs and HAs.

We should notice that the use of the MAP does not depend on, or assume the presence of, a permanent HA. On the other hand, an MN need not have a permanent HoA or HA in order to be HMIPv6-aware or use the features in this specification. An MN may use a MAP in a nomadic manner to acquire mobility management within a local domain. Section 6.5 describes such a scenario.

2.5.2 FMIPv6

The combined handover latency is often sufficient to affect real-time applications, which is suitable especially in aeronautical environment. Throughput-sensitive applications can also benefit from reducing this latency. *Fast Handovers for MIPv6* (*FMIPv6*) [RFC 4068] describes a protocol to reduce the handover latency. It is a further enhancement to MIPv6.

FMIPv6 attempts to reduce the chance of packet loss through low-latency handoffs. It attempts to optimize handovers by obtaining information for a new AR before disconnecting from the previous AR. ARs request information from other ARs to acquire neighborhood information that will facilitate handover. Once the new AR is selected, a tunnel is established between the old and new routers. The previous care-of address is bound to a new care-of address so that data may be tunneled from the previous AR to the new AR during handover. Combining HMIPv6 and FMIPv6 would contribute to improved MIPv6 performance, but this comes at the cost of increased complexity.

Two requirements are considered when developing an FMIPv6 protocol. One is how we can permit an MN to send packets when it detects a new subnet link. The other is how we can deliver packets to an MN when its attachment is detected by the new AR.

The protocol defines IP protocol messages necessary for its operation in spite of link technology. It does not depend on specific link-layer features when it allows link-specific customizations. By definition, it employs handovers that interwork with mobile IP. Relying on its new AR, an MN engages in mobile IP operations including return routability (RFC3775). There are no special requirements for an MN to behave differently with respect to its standard mobile IP operations.

FMIPv6 is applicable when an MN has to perform IP-layer operations as a result of handovers, but it does not address improving the link-switching latency. It does not modify or optimize procedures related to signaling with the HA of an MN. Indeed, while targeted for MIPv6, it could be used with any mechanism that allows communication to continue despite movements.

2.5.3 PMIPv6

In MIPv6, as described earlier, the MN updates the HA with BU messages. This mode of operation is called node-based mobility management. A complimentary approach is for access network service providers to provide network-based mobility management using PMIPv6 on access links that support or emulate a point-to-point delivery. Except from the simulation of MIPv6 by the aforementioned tool NS-3 (see Section 2.3.3), Hyon-Young Choi has also given a report on how to program an NS-3-based PMIPv6 simulation environment.[8] He also created a site for sharing his codes (https://sites.google.com/site/pmip6ns/).

MIPv6's approach that supports mobility does not require the MN to be involved in the exchange of signaling messages between itself and the HA to potentially optimize the access network service provision. A proxy mobility agent in the network performs the signaling with the HA and does the mobility management on behalf of the MN attached to the network. The core functional entities for PMIPv6 are the local mobility anchor (LMA) and the mobile access gateway (MAG). The LMA is responsible for maintaining the MN's reachability state and is the topological anchor point for the MN's home network prefix(es).

The MAG is the entity that performs the mobility management on behalf of an MN, and it resides on the access link where the MN is anchored. The MAG is responsible for detecting the MN's movements to and from the access link and for initiating binding registrations to the MN's LMA. An access network that supports network-based mobility would be agnostic to the capability in the IPv6 stack of the nodes that it serves. IP mobility for nodes that have mobile IP client functionality in the IPv6 stack as well as those nodes that do not would be supported by enabling PMIPv6 protocol functionality in the network. The advantages of PMIPv6 are reuse of HA functionality and the messages/format used in mobility signaling, and common HA would serve as the mobility agent for all types of IPv6 nodes. PMIPv6 like HMIPv6 is a local mobility management approach that further reduces the air-ground signaling overhead.

The detailed information of IP mobility for IPv6 hosts is described in MIPv6 (RFC3775). MIPv6 needs client functionality in the IPv6 stack of an MN. We can interchange signaling messages between the MN and HA to enable the creation and maintenance of a binding between the MN's HoA and its CoA. In order to realize mobility, according to RFC3775, the IP host needs to transmit IP mobility management signaling messages to the HA, which is located in the network.

There is another approach to solving the IP mobility challenge, namely network-based mobility. It is possible to support mobility for IPv6 nodes without host involvement by extending MIPv6 (RFC3775) signaling messages between a network node and an HA. This approach, which supports mobility, does not need the MN to be involved in the exchange of signaling messages between itself and the HA. A proxy mobility agent in the network performs the signaling with the HA and does the mobility management for the MN attached to the network. This protocol is known as Proxy Mobile IPv6 (PMIPv6) just for the use and extension of MIPv6 signaling and HA functionality.

Network deployments for support mobility would be agnostic to the capability in the IPv6 stack of the nodes that it serves. Enabling PMIPv6 protocol functionality in the network supports. IP mobility for nodes that have mobile IP client functionality in the IPv6 stack and those nodes that do not. The merits of developing an MIPv6-based mobility protocol are as follows:

1. Reuse of HA functionality and the messages/format used in mobility signaling. The protocol of MIPv6 is mature and many implementations have been subjected to interoperability testing.

2. All types of IPv6 nodes would have a common HA which serve as its mobility agent.

Readers interested in a network-based mobility protocol solution may refer to RFC4830 for further documentation.

The network-based mobility management protocol called Proxy Mobile IPv6 (PMIPv6) is a suggested solution being actively standardized by the IETF NetExt Working Group and is starting to attract much attention among the telecommunication community and Internet community. Unlike the various existing protocols for IP mobility management such as MIPv6, which are host-based approaches, a network-based approach such as PMIPv6 has salient features and is expected to expedite the real deployment of IP mobility management.

The fundamental foundation of PMIPv6 is based on MIPv6 in the sense that it extends MIPv6 signaling and reuses many concepts such as the HA functionality. However, PMIPv6 is designed to provide network-based mobility management support to an MN in a topologically localized domain. Therefore, an MN is exempted from participation in any mobility-related signaling, and the proxy mobility agent in the serving network performs mobility-related signaling on behalf of the MN. Once an MN enters its PMIPv6 domain and performs access authentication, the serving network ensures that the MN is always on its home network and can obtain its HoA on any access network. That is, the serving network assigns a unique home network prefix to each MN, and conceptually this prefix always follows the MN wherever it moves within an PMIPv6 domain. From the perspective of the MN, the entire PMIPv6 domain appears as its home network. Accordingly, it is needless (or impossible) to configure the CoA at the MN.

Despite the fact that network-based mobility management is considered a better solution than host-based solution, the progress of completing the detailed scheme is quite late. Moreover, there are some problems about flow mobility support.

Enhanced flow mobility support is based on the virtual interface in the MN. Virtual interface hides all physical interfaces from the network layer and above. A flow interface manager is placed at the virtual interface, and a flow binding manager in the LMA is paired with it. They manage the flow bindings and are used to select proper access technology to send the packet. Flow mobility procedure is divided into three cases, which are caused by new connection from the MN, the LMA's decision, and the MN's decision, respectively. An HNP update request/acknowledge message is defined for LMA decision-based flow mobility. According to the concept of network-based flow mobility, MN-derived flow handover needs the approval of the LMA.

2.6 NEMO and NEMO Route Optimization Analysis

Providing transparent and unperturbed Internet service is important to mobile hosts and has been studied in the IETF for many years until now. A common scenario we can easily imagine is the emergence of mobile networks, namely, a set of hosts that move conjointly as a unit, such as on flying aircrafts or any other moving vehicles. The protocols for mobility support therefore need to be extended from supporting an individual mobile device to supporting an entire mobile network. Bauer has carefully compared several technologies that support IP mobility and pointed out that NEMO is the most suitable protocol for ATN/IPS mobility.[9]

In this section, we discuss the state-of-the-art NEMO support and also give some research achievements published in recent years, especially focused on ATN environments. First, we give some basic knowledge of NEMO and then motivate the problem by considering typical aeronautical/space scenarios and identify the characteristics that require new solutions. We then study the design requirements of the protocols that support NEMO. After that, we review several current approaches for NEMO support and discuss their advantages and weaknesses (i.e., RO) in addressing the design requirements.

2.6.1 Introduction of NEMO and Its Relationship with MIPv6

The growing use of IP devices in portable applications has created the demand for mobility support for entire networks of IP devices. NEMO solves this problem by extending mobile IP. Devices on a mobile network are unaware of their network's mobility; however, they are provided with uninterrupted Internet access even when the network changes its attachment point to the Internet. IETF has already created a set of NEMO protocols to provide basic NEMO functionality on both IPv4 and IPv6 and continues to work with other projects to develop advanced performance and functionality-enhancing features. A first set of NEMO implementations are already available on several platforms including BSD variants, Linux, and Cisco Systems routing equipment. (The simulator NS-3 has already given a NEMO simulation solution; see Section 2.3.3 for details.)

Figure 2.7 depicts an example of a NEMO scenario. The router within the NEMO that connects to the Internet is called the *mobile router* (MR). (In ATN/IPS, the MR is usually mounted in the aircraft.) NEMO should reside in its home network when it is not traveling in any other foreign networks. Several addresses blocks are allocated by the home network, and then the address of the mobile network is assigned to the address block which is known as the mobile network prefix (MNP). When NEMO is away from home network, it should keep the assigned address. These addresses will lose the means of network topological structure when they are not in home network. When some data packages are being sent to a node in NEMO (this node is named by mobile network node, MNN), and the NEMO is in foreign network, at this time, the packages will be first sent to home network. MNN can be any host that always moves with other MNNs or a MR and forms a whole. The MR will exchange a prefix table with HA and will work like any mobile IP host. Based on the exchanged prefix, the HA will forward received packages to MNNs that have the same common prefix. We should note that the MR needs to be assigned an address by its visited foreign network, i.e., CoA which has been discussed before. This architecture can be used to transmit and receive packages between the MNN and CN.

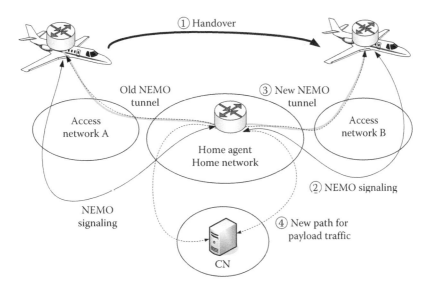

FIGURE 2.7
NEMO handover signaling between mobile router and home agent.

As described by Bernardos et al.[10], the CN can be treated as a host that is situated in any position of a network. Assume it will communicate with an MNN by some IP data packages, the signal flow can be described as follows.

- A IP data package will be transmitted from CN to MNN(A). The IPv6 address of MNN(A) is affiliated with the MNP of NEMO and is contained in the header of this IP data package.

- As mentioned above, the IP data package will be first be routed to the home network of NEMO rather than be directly transmitted to MNN(A). In the home network, a particular node called HA will repackage this original IP data package to another new IP data package. Then the new package will be sent to the MR according to its CoA; in this stage, the source address is the HA. An important point is, from the perspective of CN, that NEMO is always in its home network, i.e., the mobility of NEMO is transparent to MNN(A) and CN, and already established connections will not be corrupted while MNN(A) is moving. Most routers have the function of ingress/egress filtering, so repackaging is necessary to prevent the situation in which the source IPv6 address is incorrect or is incompatible with network topology and abandoned by the router.

- After the MR receives the repackaged IP data, it will peel the outer header, then forward the inner IP data package to MNN(A)

- In the reverse transfer stage, i.e., from MNN(A) to CN, the IP data package generated by MNN(A) is first packaged by MR and is transmitted to the HA of NEMO. Secondly, the HA will redirect the IP data package to CN.

In order to distinguish how different types of nodes move in the network topology, MNNs can be divided into three different categories, as described in the following

- Local fixed node (LFN): This type of node does not change its point of attachment in the mobile network while maintaining ongoing sessions. The LFN can be a host or a router and obtains its IPv6 address from an MNP of NEMO.

- Local mobile node (LMN): This type of node is located in the home network, but the address of the home network belongs to the mobile network. The LMN can change its point of attachment while maintaining ongoing sessions.

- Visiting mobile node (VMN): This type of node has its own home network and implementation of MIPv6 protocol, and it can be an MN or MR. Thus, maintaining ongoing sessions will not affect the VMN's attached point. A VMN can "visit" a foreign mobile network, which means the VMN can obtain an address from an MNP.

In a practical environment, a common scenario known as *nest NEMO* occurs when a mobile network is located in another mobile network. It often has the topology that an MR attaches to an MNN in another mobile network, as described by Bernardos et al. An example in civil aviation is an airline passenger connected to the in-plane Wi-Fi by his laptop through MR2 in the ceiling, while MR2 is also treated as an MNN because it is connected to the wireless aeronautical data link, which transmits all flight data (e.g., engine sensors) to the ground station.

Readers are encouraged to Review the work of Bernardos et al.[10] for more details and illustrations.

On the ingress interface (to the mobile network), the MR will advertise one or more prefixes to the MNNs. Although on the surface NEMO appears to be a straightforward extension to mobile IP, there are several considerations that are still being investigated in IETF working groups. These issues include how to do NEMO route optimization and several considerations with prefix delegation and management as it can be seen from Figure 2.7.

One possible implementation of NEMO mimics PMIPv6 in that it is the network access points that implement the mobility support. When an MN attaches to the access link, the ground access point will set up a virtual MR that registers the relevant network prefixes with the HA. All traffic received by this virtual MR for the registered MNPs will be forwarded across the air–ground link to the connected MN. The MN need not support any MIPv6 or NEMO signaling and can also auto-configure the IPv6 address if the virtual MR sends router advertisements for the MNPs across the air–ground link. When the MN connects to another ground access point, the virtual router is moved by the ground access network to the new access point with updates to the HA as if the router was physically present on the MN.

This implementation model allows for MNs with IPv6 stacks that are agnostic of the fact that they reside on a mobile link and also reduce the protocol overhead for each message that is sent across the air–ground link.

The mobility signaling for NEMO is similar to MIPv6. As soon as the MR moves to a foreign network and acquires a CoA, a home registration is performed with the HA that is located in the home network. The signaling exchange is performed by means of a BU/BA exchange, protected by an IPsec security association.

Traffic originating from the MNNs is tunneled by the MR to the HA that forwards packets to their destination, the CNs.

The home network aggregates the MNPs of all its MRs and advertises this aggregate prefix to other ASs using an interdomain routing protocol. Hence packets originating from a CN addressed to an MNN that is attached to an MR are routed to the HA located within

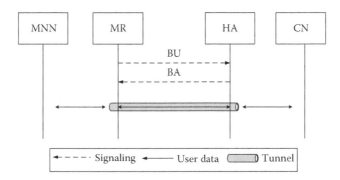

FIGURE 2.8
The signaling of NEMO.

the home network. These packets are then tunneled by the HA to the current CoA of the MR. The MR decapsulates the packet and delivers it to the destination, the MNN.

The mobility is transparent to both the MNNs and the CNs as only MR and HA are performing mobility-related tasks. The original packets are also not modified, due to the IP-in-IP encapsulation that only adds an additional header on the path between MR and HA.

Finally in Figure 2.8, we give a brief illustration of NEMO mobility signaling and forwarding of user traffic over bidirectional tunnel. In Figure 2.8, BU and BA signaling are exchanged between the MR and the HA. User traffic is exchanged between the MNN and the CN.

2.6.2 Motivation and Scenarios of Using NEMO in Aeronautical Environment

ATN specification lists NEMO in accordance with RFC 3963 (NEMO Basic Support Protocol) as an option in ATN/IPS for implementation. The motivation is MIPv6 supports movement of an individual network node. However, there are scenarios in which it would be necessary to support movement of an entire network in ATN/IPS.

As we know, several NEMO RO mechanisms already exist. In order to encourage the development of desired functions in aeronautical NEMO RO solutions, consider that aeronautical communication is often used in a terrestrial air-ground network or satellite-based ocean or polar region communication, so some network component such as access network, IIA, CN should be used and analyzed by their topology to determine which type of RO mechanism is most suitable.

- For airline passenger communications (APC) where is would be wasteful of bandwidth to have mobility signaling for every individual passenger.
- When there is a common airborne router supporting multiple traffic types provided proper security issues can be addressed.

Here are some reasons and analysis to explain why NEMO is selected as the mobility support protocol in ATN/IPS environment.

Session continuity: The HA provides the airborne router with a constant IP address—called the home address (HoA)—from its own network. Traffic is therefore always routed via the HA that forwards it to the current location of the MN. Mobile network support: NEMO extends MIPv6 by introducing an MR that has one or several mobile network.

Prefixes (MNPs). These prefixes belong topologically to the home network. End-systems that attach to the MR configure their addresses based on the MNP advertised by the MR and can therefore remain mobility agnostic.

Multihoming: The possibility to register several CoAs with the HA is specified.[11] In addition,[12] specifies a policy exchange protocol that can be used to set up forwarding rules for certain traffic flows, taking into account the additional CoA. Detailed traffic selectors can be used to identify a flow, based on IP or higher-layer protocol fields such as source/destination address, port numbers, etc. The MR sends its current policy to the HA, which sets up forwarding to the MR accordingly. This allows for simultaneously routing traffic flows over different interface, on the routing path from the MR to the HA as well as on the path from the HA to the MR.

Security: MR and HA perform a mutual authentication between each other based on IKEv2. This ensures that the HA forwards packets only to the valid MR.

End-to-end delay: NEMO causes suboptimal routing where traffic always traverses the HA. If the distance between MR and HA increases, the overall end-to-end latency also increases.

Scalability: The MR signals its current location to the HA that updates its routing state accordingly. BGP routing tables remain unchanged as the home network is always advertising an aggregated prefix via BGP that includes all the MNPs. Scalability is therefore linear with the number of aggregates, with regard to the number of announced prefixes.

Applicability to AAC/APC: The protocol stack on end-systems remains unaffected as traffic is transparently tunneled between MR and HA.

Convergence time is equal to the time it takes the MR to signal the new location to the HA, who will then immediately forward traffic to the MN's new CoA.

Efficiency 1: It takes the MR 1 RTT to signal the new location/CoA to the HA.

Efficiency 2: The tunnel between the MR and the HA inflicts an overhead of a full IP header upon every payload packet.

Support for ground-initiated communications: As it was the case for the IPsec-based approach, payload traffic is always routed via the HA. As the MR signals its CoA(s) to the HA, this traffic can be forwarded to the MR's current location.

2.6.3 NEMO Route Optimization and Its Application on ATN

A disadvantage of the current NEMO protocol is that all traffic between the MNNs and the CNs suffers from suboptimal routing, since all packets are forwarded through the HA that is located in the home network. This is in contrast to MIPv6, which does provide an RO component for establishing a direct routing path between mobile host and CN.

A significant number of related works[13,14] that have proposed RO solutions for NEMO are available. An evaluation based on a set of requirements is performed in order to select the most suitable protocol for the safety-related aeronautical communications environment. RFC5522[15] describes the requirements and desired properties of RO techniques for use in global-networked communications systems for aeronautics and space exploration. (RFC5522 is being developed by aeronautical communications experts who are outside of IETF, such as ICAO, some independent research institutes, or aeronautical communication standard groups. The latest version of RFC5522 is October 2009.) The requirements are as follows:

1. End-to-end latency: the number of intermediate nodes on the routing path between the end-systems in the mobile network and the ground network should be as small as possible.

2. Single point of failure: a new mobility-specific node for the purpose of RO should not constitute a single point of failure.

3. Separability: it should be possible to apply RO only to traffic flows that really require it. For example, RO should only be applied to all traffic originating from or destined to one specific MNN, while traffic from other MNNs is still routed via the HA.

4. Multihoming: the RO mechanism must be fully usable if several interfaces are present. More precisely, it should be possible to forward a route-optimized traffic flow via a particular interface/access network.

5. Efficient signaling: both the size and the number of individual RO signaling messages that are exchanged over the wireless path should be kept small.

6. Ground network impact: the amount of support necessary from the ground network in order to provide RO between the MNNs and the CNs should be limited.

7. Security: the mobility entity on the ground with whom RO signaling is performed must be able to validate the aircraft as proper owner.

8. Adaptability: the RO scheme should not break applications using new transport protocols or IPsec.

This set of requirements used in evaluation is partially overlapping with the set that was used for deriving NEMO as the most suitable mobility protocol. The requirement on applicability to Aeronautical Administrative Communication (AAC) and APC has been omitted as this investigation is focused on supporting safety-related services. The requirements on routing scalability, convergence time, and support for ground-initiated communications have been omitted as they are fulfilled by NEMO, irrespective of the selected RO protocol. Additional NEMO-specific requirements have been added though, such as separability or adaptability.

In Bauer's research,[14] various different proposals for performing NEMO RO will be assessed in the following based on the requirements as mentioned before. The RO procedure can be initiated and performed by different network nodes. Consecutively, four types of RO strategies are defined by the entities participating in RO signaling. These strategies are shown in Figure 2.9.

- MNN to CN
- MR to CN
- MR to correspondent router
- MR to HA

In Table 3.1 of Bauer's PhD thesis,[14] a summary of the assessment of the four NEMO RO solution classes with regard to the previously specified requirements is provided.

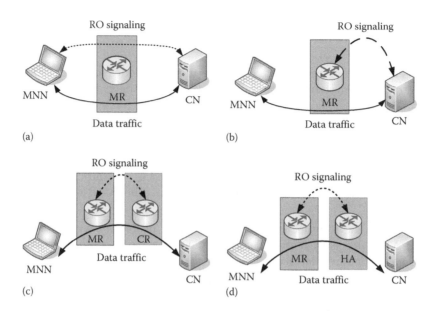

FIGURE 2.9
Four different NEMO route optimization classes. (a) MNN to CN, (b) MR to CN, (c) MR to CR, and (d) MR to HA. (Adapted from Bauer, C., *Secure and Efficient IP Mobility Support for Aeronautical Communications*. Karlsruhe, Germany: KIT Scientific Publishing, 2013.)

To better help readers to understand the analysis process better, here we give a brief review of evaluation of the four optimization types of NEMO. For more details, please refer to Bauer[14].

1. *MNN to CN*: The MR only relays the router advertisement of its own access router(s) to the MNNs. Then the MNNs make use of their own Mobile IPv6 functionality to form a CoA from this advertisement. Then they perform the Mobile IPv6 RO signaling with the CNs themselves. So in this method, The MR just plays the role of a supporter.

   a. *Advantage*: Optimized path is the shortest possible one with direct routing between MNN and CN. It does not have the problem of single point failure. MNNs can easily decide for which data flows RO should be performed. Multihoming is also supported

   b. *Disadvantage*: Overhead is heavy, as RO signaling is performed between every MNN and CN for which an optimized routing path should be provided. The MN will also have to manage the CoA of the MNN.

2. *MR to CN*: MN performs the MIPv6 RO signaling on behalf of the MNN with a CN. The MR therefore acts as a RO proxy for the MN.

   a. *Advantage*: The end-to-end path is still optimal because packets are being routed directly on the path between MR and CN. No single point of failure is introduced. The MR can also make use of multihoming by registering several CoA with a particular CN.

   b. *Disadvantage*: Security weaknesses. The MR also has to extend the original packets with additional headers when performing RO on behalf of the MNN, so it increases the overhead. The integrity of end-to-end flow is corrupted, so some security protocols may take the RO signal as a attack.

3. *MR to CR*: A correspondent router (CR) is newly introduced in this method. It is located close to the CNs. The MR exchanges mobility signaling with the CR and constructs a bidirectional tunnel that is used to exchange traffic between the mobile network and the network served by the CR. The CR therefore acts as a RO proxy for the CNs.

   a. *Advantage*: If there is no signal global CR, the problem of single point failure does not exist. Also the signaling overhead has been reduced compared to the previous protocols (1) and (2). This RO process is transparent to end-to-end transport and security protocols

   b. *Disadvantage*: Some security risks are inherited, since the signaling procedure between MR and CR is based on MIPv6 RO signaling.

4. *MR to HA*: The concept of having only a single HA has been extended to having multiple, distributed HAs within the home network. The proposal is called Global HA-to-HA. The MR binds to the closest HA to achieve RO. The amount of end-to-end latency reduction depends on the location of the HA.

   a. *Advantage*: The signaling exchanges between MR and primary HA are similar to the basic NEMO protocol. Global HA-to-HA does not require any mobility functionality in the CNs or within the CN networks. It not only provides a high level of security between MR and HA by IP-in-IP tunnel, but also the end-to-end transport or security protocols remain unaffected.

   b. *Disadvantage*: The single point of failure problem has only been partially addressed. The amount of end-to-end latency reduction depends on the location of the home agents.

As a conclusion, the above methods (1) and (2) are discarded because the complexity is increased at the end-systems, and also because of large signaling overhead as well as security risks. Both (3) and (4), illustrated in Figure 2.10, have their pros and cons. The CR protocol in (3) is the most adequate solution for safety related aeronautical communications. For non-safety related communications, especially aeronautical passenger communications, protocol (4) is the only feasible solution.

2.7 Conclusion

As we know the essential characteristics of aircraft are high speed and high mobility, ATN should conform to these characteristics. How to keep accurate and timely data transmission is a critical aspect of the ATN system. MIPv6 is the solution for these problems in the IP layer.

In this chapter, we give an overview of MIPv6 and its extension, which can be used in the next-generation ATN system. The discussion is started from the infrastructure of general mobile communication, then the motivation of MIPv6, MIPv6 protocol, and related triangular routing problem is presented. Due to some defects of the original MIPv6 protocol, four types of extension are introduced, that is, HMIPv6, FMIPv6, PMIPv6, and NEMO.

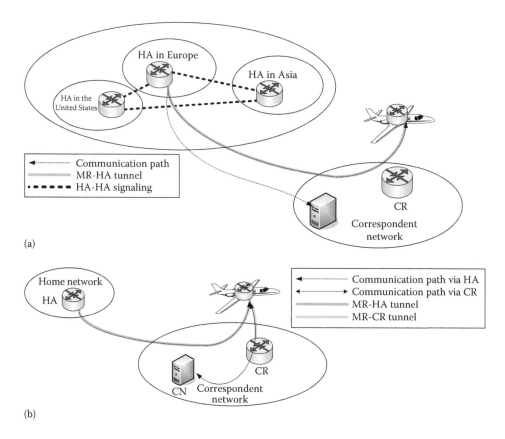

FIGURE 2.10

(See color insert.) Typical air traffic services communication scenario making use of different route optimization protocols. (a) Global HA-to-HA and (b) communication router. (Adapted from Bauer, C., *Secure and Efficient IP Mobility Support for Aeronautical Communications.* Karlsruhe, Germany: KIT Scientific Publishing, 2013.)

Since NEMO is very suitable for a whole mobile network moving, we focus on optimization of NEMO and its applications.

This chapter is based on the recent published research articles and technical reports; the readers can catch the essence of mobility of ATN and can conduct deep study about the knowledge of a certain subsection.

References

1. ICAO. Aeronautical Communications Panel (Working Group I), Manual for the ATN using IPS standards and protocols, Doc 9896, First Edition, April 2014 (More details about Working Groups of ACP Panel are available at http://www.icao.int/safety/acp/Pages/).
2. Sampigethaya, K. Future E-enabled aircraft communications and security: The next 20 years and beyond, *Proceedings of the IEEE*, 99(11), 2040–2055, 2011.
3. EUROCONTROL. Challenges of growth 2013, Task 7 Report: European air traffic in 2050. Available at https://www.eurocontrol.int/articles/challenges-growth. June, 2013.

4. Fistas, N., and B. Phillips. Action plan 17—Future communications study—Final conclusions and recommendations report. *Digital Avionics Systems Conference (DASC) Proceedings*, 2007, p.12.

5. EUROCONROL. Communications operating concept and requirements for the future radio system (COCR V1.0), Eurocontrol/FAA Report, 177pp.

6. Sajatovic, M., Haindl, B., Epple, U., and Gräupl, T. Updated LDACS1 system specification. SESAR P15, 2, 2011. More infomation can be found at http://www.ldacs.com/publications-and-links/.

7. Microsoft Corporation. Understanding mobile IPv6. Microsoft Windows Server 2003 white paper, 2004. http://www.cu.ipv6tf.org/pdf/MobileIPv6.pdf pp 5–7.

8. Choi, H.-Y., Min, S.-G., and Han, Y.-H, PMIPv6-based flow mobility simulation in NS-3, *Fifth International Conference on Innovative Mobile and Internet Services in Ubiquitous Computing* (IMIS), pp. 475–480, June 30, 2011–July 2, 2011.

9. Bauer, C. and Zitterbart, M. A survey of protocols to support IP mobility in aeronautical communications, *IEEE Communications Surveys and Tutorials*, 13(4), 642–657, Fourth Quarter 2011.

10. Bernardos, C.J., Soto, I., and Calderón, M. IPv6 network mobility, *Internet Protocol Journal* 10, 2, 2007. http://www.cisco.com/web/about/ac123/ac147/archived_issues/ipj_10-2/102_ipv6.html.

11. Wakikawa, R., Devarapalli, V., Tsirtsis, G., Ernst, T., and Nagami, K. Multiple care-of addresses registration, RFC 5648, October 2009.

12. Tsirtsis, G. et al. Flow bindings in mobile IPv6 and network mobility (NEMO) basic support. IETF request for comments 6089 (2011). Details can be found at https://tools.ietf.org/html/rfc6089.

13. Bauer, C. A secure correspondent router protocol for NEMO route optimization, *Computer Networks*, 57(5), 1078–1100, April 2013.

14. Bauer, C. *Secure and Efficient IP Mobility Support for Aeronautical Communications*. Karlsruhe, Germany: KIT Scientific Publishing, 2013.

15. Eddy, W., Davis, T., and Ivancic, W. Network mobility route optimization requirements for operational use in aeronautics and space exploration mobile networks (RFC 5522). October 2009. More details can be found at http://tools.ietf.org/html/rfc5522.

3

Modern Air Traffic Control Systems

Yun-Fei Jia

CONTENTS

3.1 Introduction

Airplanes must be safely guided in the sky and on the ground. The objective of the air traffic control (ATC) system is to make sure airplanes be safely separated as they fly in the sky, land at or take off from the airport, or move around the airport. In addition, ATC can enhance the airspace usage and expedite the flow of traffic. This chapter introduces the ATC system.

The ATC system is the man–machine interface that provides various services to in-flight airplanes. In many countries, the ATC system is set up at air traffic control centers (ACC), where controllers issue instructions and advisories to pilots.

ATC is a distributed computer system, including several subsystems. These subsystems process the radar data, calculate the minimum separation between airplanes, manage the flight plan, and display the tracks. Usually, each subsystem is running at individual computers, and all the computers are connected via a switch, forming a distributed computing system. These computer systems are connected by a dual network. This design can prevent network failure. In addition, each computer runs a real-time operating system. This guarantees the ATC's response to the request of a pilot within a certain period of time.

This chapter focuses on the operational principle of the ATC system. We first provide an overview and discuss the evolution of ATC, including its composition and how it works. In Section 3.3, we describe the radar data processing (RDP) subsystem, including interpretation of the radar data package, ASTERIX protocol. This section focuses on the process of track generation. The conflict detection and warning (CDW) subsystem is described in Section 3.4, including short-term conflict and warning, minimum safety altitude warning, dangerous area infringement warning, and special code warning. Section 3.4 elaborates the principle of the design of CDW. Flight planning is the core part of ATC, as is illustrated in Section 3.5. State management of the flight plan is very important for ATC. The flight profile, which is calculated from flight plan, can estimate the quality of flight. Flight plan handover is also mentioned in Section 3.5. The Integral Surveillance and Control (ISC) terminal is a man–machine interactive system. It shows the calculated results of all the subsystems, including tracks, conflict warning, and flight plan. All these will be described in Section 3.6. In addition, record and replay is an important part of the ATC system, the operating mechanism of which is also described. Section 3.7 concludes this chapter.

To sum up, this chapter will make you aware of how ATC works to guide an airplane safely to its destination airport. The operational principles of the ATC system are also included.

3.2 Overview of ATC

3.2.1 Evolution of ATC

With the fast development of commercial aviation, air traffic congestion becomes more and more serious. In order to guarantee the safety of more and more flying airplanes and improve the efficiency of airspace usage, the ATC system is designed to guide airplanes according to a set of predefined rules. Air traffic control refers to the activity of controlling and/or guiding the airplanes so that they can fly to their expected destinations safely.

ATC has undergone four stages of development, which can be roughly described as follows: Before the 1930s, airplanes flew only in daytime, and the pilots obeyed a set of rules for "visual flight." Not long afterward, controllers were employed to guide the landing and taking off of airplanes with red and green flags. Also, a control tower was set up at the highest point of the airport, where the controller could monitor the airplanes on the ground through visual means. From 1934 to 1945, visual flight rules hardly met the requirement of increasing flight activities. During this period, radio communication and navigation devices were mounted on airplanes and around the airport. In addition, some air traffic administrations such as towers, local ACCs, and flight stations were set up across the country. More specifically, towers can guide the movement on the airport ground, monitor departure and arrival, and local control centers can guide airplanes as they are cruising on the air route. The local control centers are responsible for collecting the position of airplanes, and the controller can determine the relative position of airplanes and thus direct the flight by communicating with the pilots. From 1945 to 1980s, radar technology, which was developed during World War II, was widely used for ATC. The controller can see on the screen the position, call number, height, speed, and related parameters of in-flight airplanes. Another important technical advancement is the instrument landing system (ILS), which enables airplanes to land in bad weather. The fourth stage began in the 1980s, where satellite communication technology and the Internet are being extensively applied to ATC. With the help of satellites, the ATC system is united with the airplanes in the air, thus processing communication automatically.

3.2.2 ATC Overview

The primary task of ATC lies in controlling and managing the whole process from take off to landing. In this process, ATC must guarantee that the airplane will not conflict with another airplane, or with barriers on the ground. Also, ATC must expedite air traffic and enhance airspace usage [1,2]. Classified by the controlled airspace, ATC can be divided into airport control, approach control, and local control.

3.2.2.1 Airport Control and Approach Control

Airport control is usually responsible for guiding the landing process, movement on the ground, and departure. The controllers in the tower manage and control the airplanes on the ground mainly by visual means. A tower is a tall building with many windows located on the airport grounds. Tower controllers are responsible for the separation and efficient movement of airplane and vehicles on the runways. In addition, some busier airports may use surface movement radar to help the controller to efficiently manage airplanes. Consequently, there are usually several displays in the control tower, which can help the controller to guide the airplanes on the ground. The display may include a map of the local area, the position of various airplanes, meteorology information, and so on.

Approach control intends to guide incoming cruising airplanes to gradually lose altitude for landing, and outgoing airplanes to gain altitude until they reach the cruising altitude. Usually, airport control and approach control are carried out by different controllers, yet in the same tower. Only in a busier airport is approach control part of individual ACC.

Approach control employs radar to monitor the airplane and guide it via radio communication. It uses the same surveillance technology as local ATC.

3.2.2.2 Local Control

After an airplane takes off and climbs to the cruising altitude, local control centers will take care of it. Because airplanes often fly across a wide area, there are several local control centers set along the air route. Each will serve the airplane in their respective airspace. According to the flight plan of an airplane, the controllers in the local control center will either accept or reject the airplane handed over from the previous local control center. Then they guide the airplane via radio communication and radar, so that there is a safe distance between airplanes. Finally, the controller will hand over the airplane to the next local control center. This process is repeated until the airplane reaches its destination airport.

3.2.3 Components of ATC

The ATC system is a distributed computer system. Each subsystem runs on an individual computer. These computers are connected in a LAN via a switch. All these subsystems are connected via TCP/IP protocol. Figure 3.1 illustrates the structure of an ATC system.

The flight plan subsystem will receive and then store in database the flight plan from airlines. ISC can obtain these plans by querying the database.

The surveillance data from primary radar, secondary surveillance radar (SSR), or ADS-B are sent to this LAN. The first receiver is the radar data processing subsystem, which interprets the radar data and extracts trace points from them. Then, it translates the collected trace points to tracks. This work is completed by determining whether the trace points represent the new positions of the existing track or a new aircraft.

RDP outputs the tracks that include, but are not limited to, an ID and positional information about an airplane. Then the tracks are sent to the CDW subsystem, where they are compared against neighboring airplanes or barriers on the ground to decide whether an alarm should be triggered. Then the CDW subsystem sends these tracks to the ISC subsystem, which is a GUI application. The tracks will be shown on the ISC.

In addition, the reliability and availability of an ATC system is very important. Hence, double-network structure is usually employed to avoid network failure, as is shown in Figure 3.1. Certainly, a highly reliable and available system should adopt several fault-tolerance techniques. However, this chapter is intended to illustrate the operational principle of ISC instead of reliability design of ATC.

All the subsystems are elaborated in the following sections.

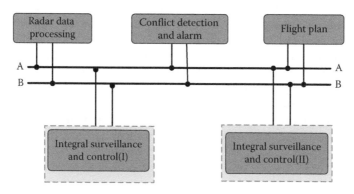

FIGURE 3.1
(See color insert.) Composition of ATC.

3.3 Radar Data Processing

Radar data processing (RDP) will receive data collected by several radars and transmit them to a uniform target track usually by means of data fusion. This process includes interpretation of a radar data package (may be from various radar manufacturers).

3.3.1 Format of Radar Data

3.3.1.1 Radar Data Transfer Protocol

Transmission protocol for radar data belongs to the data link protocol in the OSI reference model, including asynchronous protocol and synchronous protocol. Synchronous protocol includes BISYNC and HDLC, and the latter is the de facto protocol used in the aviation industry. HDLC protocol is described as follows.

High-Level Data Link Control (HDLC) is a bit-oriented data link protocol, used to transfer radar data on a synchronous network. HDLC protocol is developed by extending IBM's Synchronous Data Link Control (SDLC) protocol. The frame structure of HDLC is illustrated in Figure 3.2.

In Figure 3.2, the F1 field denotes the starting point of the frame. Moreover, it is used to fill certain characters between two frames. Field A represents the address of the source radar. Each source station has a unique address. However, an address can represent a set of stations, which is called station set. This field can be set to 11111111 to represent the broadcast address. This field can also be set to 00000000 to be used for testing purposes. In Figure 3.2, C represents control field, which is a key field of the HDLC frame. This field is used to construct various commands and responses for the purpose of surveillance and control of data link. For example, the sender can use this field to tell the receiver to execute specified commands. Also, the receiver can use this field to reply the execution status of the instructions to the sender. The first and second bits in this field represent the type of transfer frame, including information frame, surveillance frame, and unnumbered frame. The fifth bit of this field is P/F bit, representing polling. The information field is denoted by I in Figure 3.2. This field contains the data to be sent, the size of which depends on the size of the buffer of the computer. Usually, the size is from 1000 to 2000 bits. The last field FCS represents the frame check sequence, which usually adopts a 16-bit CRC algorithm.

3.3.2 ASTERIX Protocol

Depending on the manufacturer of the primary radar, the format of radar data can be different, including ASTERIX, CD2, MP2, TOSHIBA, MH/4008-2000, etc. The most commonly used format is ASTERIX. An ASTERIX radar package may contain one or more data blocks. A data block is composed of category of data (CAT), length of data block (LEN) and a certain number of records. The data block can be described by Table 3.1.

8 bits	8 bits	8 bits	8 bits*n	16 bits	8 bits
F1	A	C	I	FCS	F2

FIGURE 3.2
Frame structure of HDLC.

TABLE 3.1

Format of ASTERIX Data

CAT	LEN	Data Record (1)	Data Record (2)	......
1 byte	1 byte	Variable-length byte	Variable-length byte	......

3.3.3 Track Generation

The primary purpose of RDP is to translate the collected trace points to tracks. A track is a time series, representing a series of history positions of an aircraft. Each track has a unique track ID. Trace point is the flying object observed by radar. After each rotation, the radar collects many trace points. Each trace point may represent a new aircraft or new position of an existing track. This process is called track generation.

In a scanning period, when receiving new radar data, RDP will usually do the following tasks as shown in Figure 3.3.

3.3.3.1 Protocol Interpretation

Because the received data may be collected by radars from different manufacturers, they usually use different transmission protocols as mentioned in Section 3.3.1. Hence, RDP first interprets the radar data package. More specifically, RDP extracts the positional information, speed, and the secondary surveillance radar (SSR) code from the radar data package.

3.3.3.2 Coordinate Transformation

Radar records the position of airplanes in polar coordinates for convenience, while the maps and air lines usually employ rectangular coordinates. Hence, we should transform the positional information, included in radar data packages, to rectangular coordinates. Moreover, the positional information included in radar data will be corrected. This is because the location, the sea level at which the radar is located, will result in error of observation. Hence, each radar data will be corrected depending on the reasons for observation error.

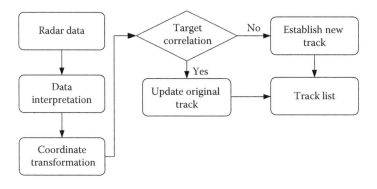

FIGURE 3.3
Track generation.

3.3.3.3 Target Correlation

In an observation cycle, radar will collect several trace points. These trace points may correspond to existing track, or a new one. If it belongs to an existing track, then it will be marked with an existing track ID. Accordingly, in the next display update cycle, ISC updates the position of an existing track with this ID. Otherwise, it will show a new track.

Target correlation aims to determine whether a new trace point corresponds to an existing airplane, or a new one. This is done by comparing the position, speed, altitude of incoming trace point against nearby existing tracks.

When target correlation is completed, the generated track will be added to the track list and sent to the CDW subsystem.

3.4 Conflict Detection and Warning

3.4.1 CDW Overview

The first task of ATC is to ensure safe flight of airplanes. The ATC system should warn the controller when there is any potential danger or harm to flight activity. Accordingly, the controller can take measures to eliminate or reduce the harm or potential risk. CDW is designed to automatically evaluate the potential riskful conditions or harm based on the surveillance information.

The CDW subsystem includes several types of warning, that is, short-term conflict warning, minimum safety altitude warning, dangerous area infringement warning, special code warning, and so on.

Conflict warning is judged by the CDW and shown on the ISC. If there is any conflict warning, various warning signals will be shown on ISC. ISC shows flickering track plates in different colors. Usually, a yellow flickering track plate denotes a less serious risk situation and red flickering track plate indicates dangerous conflicts. In addition, ISC warns the controller with a warning sound. Conflict warning is illustrated in Figure 3.4.

Once the controller gets a warning signal, they can roughly judge the cause of conflict and quickly identify the track in question. Then, according to the causes of conflict, the controller can guide the aircraft to eliminate the conflict condition. If the conflict condition is eliminated, the warning signals will gradually disappear. Typical conflict conditions are described in the following sections.

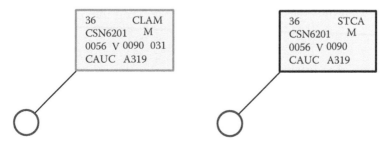

FIGURE 3.4
(See color insert.) Conflict warning.

3.4.2 Short Term Conflict Warning

Short-term conflict warning (STCW) will warn the controller of possible conflicts between airplanes. The ATC system will issue an STCW when the horizontal interval *and* the vertical space of flight-pair are less than the minimum safety distance.

STCW will calculate the interval of each flight-pair. Thus, when there are N airplanes flying in the airspace of concern, the number of flight-pairs (denoted by M) can be calculated by

$$M = C_N^2 = \frac{N \cdot (N-1)}{2}.$$

From this equation, we can obviously see that when there are too many airplanes in the airspace, the number of flight-pairs can be astronomical. This will overload the ATC system, leading to inability to output real-time warnings. For simplicity, we can use filtering methods to reduce the computing time. Typical filtering methods include mosaic filtering, coarse filtering, and fine filtering. In this book, we only introduce the mosaic filtering method.

The mosaic filtering method divides the horizontal project of the airspace into a number of mosaics, and each mosaic includes a number of tracks. Because the flying speed is a limited value, two mosaics far from each other will not result in short-term conflict. Thus, we can filter some flight-pairs for further processing.

It should be noted that the mosaic filtering method only considers the horizontal space between two tracks. Hence, the length of the side of a mosaic should be well considered. If it is too long, there will be too many tracks in a mosaic, and the filtering effect will be distorted. If the size is too small, there will be too many mosaics in the airspace of concern, resulting in poor searching performance.

3.4.3 MSAW

A minimum safety altitude warning (MSAW) is designed to detect the interval between airplane in the air and the obstacles on the ground. Obviously, this type of warning detection depends on the topographic information of the airspace of concern. For simplicity, we can adopt a square mosaic, the side length of which equals the highest point of the topography of concern, to represent the minimum safety height.

Under some conditions, MSAW should be restrained. For example, when an aircraft is taking off or landing, there is very small space between the aircraft and the buildings on ground. In this case, the false warning signal will disturb the work of the controller. MSAW can be restrained by the controller by specifying the area where MSAW is not calculated.

3.4.4 Dangerous Area Infringement Warning

A dangerous area infringement warning (DAIW) is issued when an airplane is close to or infringing some special area. These special areas include restricted airspace, dangerous airspace, and so on. Dangerous areas can be temporal or long-term. Some long-term dangerous areas are specified beforehand by the controller. Also, there are some temporal dangerous areas. For example, when there are military manoeuvres, that airspace will be restricted for a period of time. Temporal dangerous areas are also specified by the controller.

Dangerous areas are defined by a polygon column, the top surface of which is parallel to the ground. A metric denoted by the minimum safety distance is defined in advance. If the distance between track and that predefined polygon column is less than the minimum safety distance, a DAIW will be issued to the controller.

3.4.5 Special Code Warning

Besides these three types of warnings, some special warnings are also issued. In extreme cases, the pilot will press the emergency buttons on the instrument panel in the cockpit. Accordingly, the ATC will receive some special codes from the pilot, including 7500 (hijacking), 7600 (onboard radio devices failure), and 7700 (airplane in emergency). Under those conditions, the ISC will warn the controller.

3.5 Flight Plan Processing

The flight plan is a detailed description of the coming flight activity. A piece of flight plan includes airplane identifier (usually airline number and task number), flight rules (instrument flight, visual flight), flight category (general aviation flight, military flight, non-periodical flight, periodical flight), airplane number, type of airplane, type of wake (heavy, medium, or light), onboard devices, departure airport, expected departure time, cruising altitude, destination airport, expected departure time, alternate airport, secondary alternate airport, etc.

The flight plan is established by the aviation operation units (usually airlines), and sent to air traffic administration. When the ATC system receives a piece of flight plan, it will store it in its database and manage its state. Also, the received flight plan can be modified at the ISC by the controller.

Flight plans can be long term and temporal. In commercial aviation, most flight plans are relatively constant. Long-term flight plans are stored in the ATC system. Thus, the airline does not need to send a flight plan to ATC, unless it is changed.

3.5.1 State Management

Flight plan processing (FPP) will take charge of the state transition of a flight plan, including automatic acceptance of flight plan, storage of flight plan, management of state, assignment of SSR code, correlation of flight plan with track, forecasting track, and track handover.

State of flight plan includes inactive state, preactive state, coordinated state, controlled state, handed-over state, and finished state [3]. Figure 3.5 shows the state transition of a flight plan in chronological order.

When FPP receives a piece of flight plan, it sets it to the inactive state. Before a period of time the airplane takes off (usually half an hour), the flight plan will be set to the preactive state. Just before the airplane takes off, the air traffic controller will communicate with the pilot to determine if everything is OK. Then, the airplane will take off and the flight plan will transit to the coordinated state. After the airplane takes off, the radar will capture it and correlate it with its flight plan, then the flight plan will transit to the

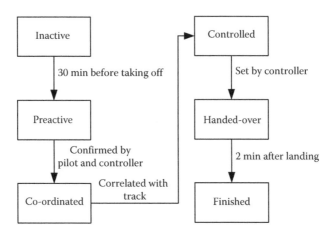

FIGURE 3.5
State transition of flight plan.

controlled state. Because an airplane usually flies a great distance, several air traffic centers are located along the flight line. When the airplane flies past an air traffic center, the controller will set the flight plan to the handed-over state. In such a situation, both the current air traffic center and the next air traffic center will see the airplanes on the screen of the ISC. After the controller at the next air traffic center accepts the flight plan, it will transit to back to the controlled state. Usually, after a period of time (usually 2 min) from the airplane's landing at destination airport, the flight plan will automatically transit to the finished state. At this point, the lifetime of the flight plan is over.

Civilian airplanes are usually identified by an SSR code, which is a 4-bits octal number. Thus, there will be $2^{12} = 4096$ SSR codes at most. In fact, some codes are used only for special purposes, so the available SSR codes for civilian airplanes are very limited. In countrywide airspace, there may be more airplanes than the limited number of SSR codes. The solution is to assign a unique SSR code to a civilian airplane in a certain control area. While the airplane flies to another control area, the assigned code will be changed. The rationale behind this solution lies in that, in a control area, there are a limited number of flying airplanes and each can be assigned with a unique code.

Another important function of FPP is to correlate track with the corresponding flight plan. This is implemented through the SSR code assigned to the airplane by the controller in a certain control area. Further, the control center in that control area will use a secondary surveillance radar to supervise the airplane, the ATC in that control center will receive the secondary surveillance radar data that contains an identical SSR code to that mentioned earlier. Thus, the track can be correlated with the flight plan.

3.5.2 Flight Profile

FPP can be used to forecast when the airplane arrives at each way point and by what route. Further, the flight profile can be estimated, which includes the time and altitude of the airplane arriving at each way point. This information is extracted from the flight plan. A possible flight profile is illustrated in Figure 3.6.

A complete flight process can be divided into three stages: climbing stage, cruise stage, and descent stage. The airplane will depart the airport and then try to climb to cruise

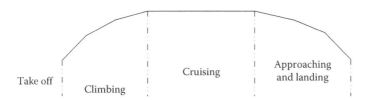

FIGURE 3.6
Flight profile.

altitude. The airplane will fly at the cruise altitude usually at constant speed. Finally, the airplane will descend steadily when it approaches the destination airport. It should be noted that, in Figure 3.6, the height of the starting point and the end point of the profile is usually different. This is because the altitude above sea level of the departure airport and the destination airport will be different. In the flight plan, we can obtain the code of departure airport and thus its height above sea level. Based on the airplane model field in the flight plan, we can estimate the climbing rate and descent rate. Also, with the expected departure time field and cruise altitude in the flight plan, we can draw the curve at the climbing stage. The curve at the cruising stage can be easily drawn, because the expected time to arrive at each way point (denoted by E1, E2, E3, and E4) can be directly obtained from the flight plan, and remember the cruise speed is constant. At the descent stage the curve can be drawn in the same way as that at the climbing stage.

Flight profile can also be used to calculate the expected flight-related parameters (flight route, speed, etc.) based on the approved flight plan, and warn or prompt the controller if the actual flight parameters are inconsistent with those in the flight plan.

In addition, a flight profile can show the planned track on the ISC. This is meaningful when the aircraft is flying in the blind zone of the primary radar. Under this condition, the track will disappear on the ISC and the planned track (denoted by a different tag) will be shown. When the aircraft has flown across the blind zone, the track will be shown again and the planned track will disappear.

3.5.3 Handover

In two cases, the FPP will hand over the flight plan. In the first case, when airplane flies across control areas, the FPP will transfer the control to the next control area. In the second case, when the airplane flies across control sectors, the FPP will transfer it to another controller. In both cases, FPP should change the state and send necessary flight plan information to the receiver. Handover is elaborated in Section 3.6.5.

3.6 Integral Surveillance and Control

ISC is a man–machine interactive system, which is an important part of the ATC system. The main functions of the ISC are threefold: (1) to display the supervision data from the primary radar, infrared camera, or ADS-B; (2) to process the flight plan and present it in the form of a flight strip; and (3) to exchange information with the controller and the pilots.

In an ATC, there may be more than one ISCs cooperating with each other to complete the task. In Figure 3.1, two ISCs are employed, that is, ISC I and ISC II.

3.6.1 Overview of ISC

The main window of the ISC is depicted in Figure 3.7, which can be roughly divided into three fields: the formation field on the top toolbar, man–machine interaction field in the middle part, and the function field at the bottom toolbar.

The features in the format field can be illustrated as follows. We can set the System mode to "NORM MODE" or "REPLAY MODE." As its name implies, NORM MODE means the ISC is currently used to monitor the air situation, while REPLAY MODE refers to the ISC being used as a terminal to replay the air situation. REPLAY MODE is usually used for accident analysis or newbie training. The measurement units are either imperial units or SI. Users can select the time zone in the UTC time field. An alarm clock can be used to remind the controller about important matters at specific time intervals. The longitude-latitude field shows the cursor position.

The man–machine interaction field lies in the middle part of the main application window. This is the main field to show and/or manipulate the tracks, flight trip, and various maps such as section maps, landmark maps, border maps, and so on. For example, users can select any flight trip and edit it, or select any track to change the style, or measure the distance between two tracks. The concentric rings shown in Figure 3.7 help the controller to determine the distance between two points.

The function field at the bottom toolbar provides most of the functions of the ISC. Clicking any button in this field opens a pop-up window with more details. Clicking the "Map" button shows each map layer. The "Flight Plan" button can be used to set or edit the flight plan in detail. The "Drawing" button is used to select drawing tools to draw maps on screen. This field also provides a record and replay function, an infrared surveillance function, etc.

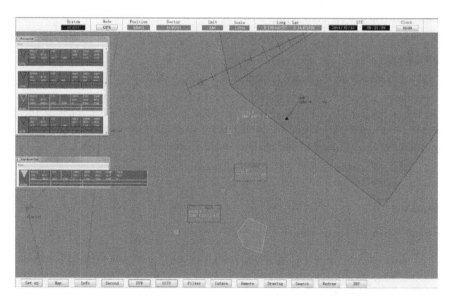

FIGURE 3.7
(See color insert.) Overview of ISC.

In this section, the five modules of the ISC will be described, that is, track and maps, flight plan, infrared surveillance, record and replay, and track handover.

3.6.2 Tracks and Maps

On the screen of ISC, a real-life airplane is denoted by a hexagon symbol. This symbol is referred to as a track in this section. It is linked with a piece of string, called a track plate. The track plate shows detailed information about the flying airplane, such as the altitude, position, international registration number, and so on. The ISC uses different symbols to represent tracks to clearly differentiate the data source. The data from radar is indicated by the symbol "+" and from ADS-B by the symbol "△." When manipulating the track, users should select the track first. When the track is selected, a red hexagon is set around it, and the corresponding flight strip is also selected. A typical track and its track plate are shown in Figure 3.8.

From Figure 3.8, we can see some dashed dots following the track that represent the historical positions of this airplane in a period of past time. The controller can optionally set how many history trace points will be shown. History trace points can help the controller to determine the airplane's point of origin.

Usually, the ISC includes three types of maps, system map, online map, and generated map. System maps refer to those relatively static maps, such as airspace map, section map, air line map, guidance station map, airport map, and so on. These maps are designed as layers. Thus, user can display any number of maps arbitrarily. If necessary, these maps drawn by the controller can be released to other ISC terminals, when two or more ISC terminals are employed, as shown in Figure 3.9. For example, temporally risky area maps are often released to other ISCs.

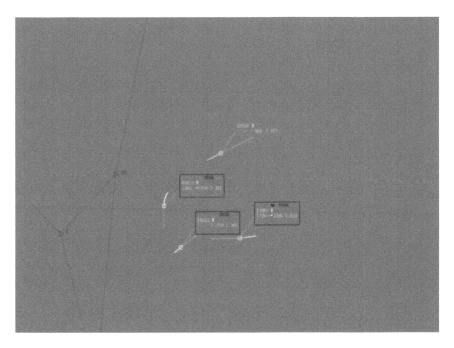

FIGURE 3.8
(See color insert.) Track.

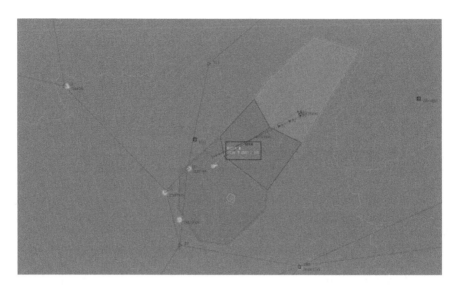

FIGURE 3.9
(See color insert.) Maps.

3.6.3 Browse and Modify Flight Plan

If these airplanes have submitted a flight plan, they will get useful information from the ATC system when they are flying, such as meteorology, military activity area, ground proximity alarm, and so on. [4]. Particularly, the airplanes flying a commercial route will get beneficial information. A flight plan window is shown in Figure 3.10.

The ISC flight plan window provides rich information about the flight process, mainly including type of airplane, expected time of departure, expected time of arrival, destination

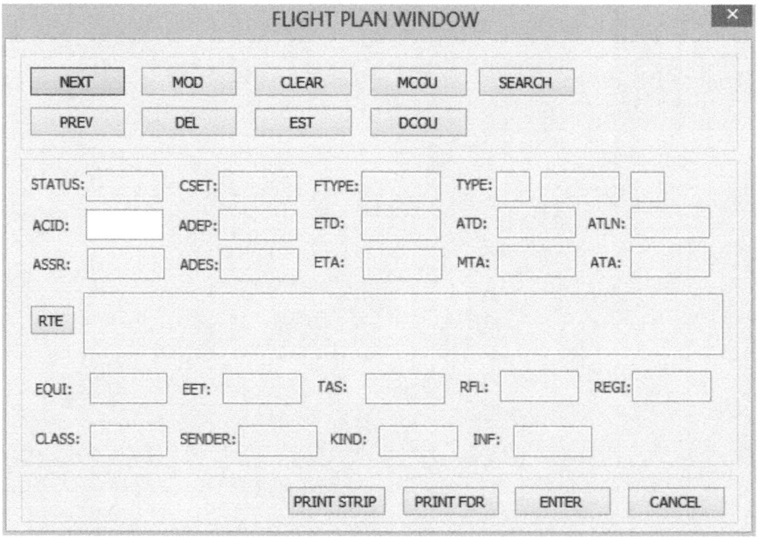

FIGURE 3.10
(See color insert.) Flight plan window.

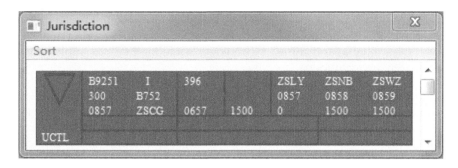

FIGURE 3.11
(See color insert.) Flight strip.

airport, etc. Moreover, if the in-flight airplane modifies its flight plan, this flight plan window can update the information in real time.

There are primarily four types of flight-plan-related manipulations: flight plan browse, flight plan modification, flight plan state change, and flight plan correlation. There are two ways to browse a portion of the flight plan: (1) select a track, then open flight plan window, which displays the flight plan of a particular track, and (2) directly input the flight number into the flight plan window to obtain the details. In some cases, an airplane might change its intention of flight when it is flying. For example, when an airplane changes its destination airport, flight route, or expected time of arrival due to bad weather, its flight plan should be modified accordingly. The controller can edit fields as shown in Figure 3.10 to modify the flight plan and synchronize it with the database. Flight plan state change means it changes to the preactive state from the inactive state, or to the active state from the pre-active state, and so on [3].

Flight plan correlation means the correspondence between the flight plan and track. It should be noted that a track plate will not be shown around the track unless the track is correlated with the flight plan. In most cases, the controller will frequently observe the flight plan. Thus, a compact version of a flight plan called the flight strip is created, which is always shown on the screen. A flight strip is illustrated in Figure 3.11.

It should be noted that all the information shown in a flight strip can be found in the flight plan window. In Figure 3.11, all fields of the flight-plan-related information are arranged in a specific order, and each field represents some information extracted from the flight plan.

3.6.4 Record and Replay

Record and replay is a powerful technique to analyze the cause of an air accident. Record and replay will record the surveillance data, flight plan, operations of the controller, etc. When air accidents occur, we can replay the air situation to analyze the cause and investigate the responsibility of the controllers. Also, the feature of record and replay is often used for training newbies [5].

The design objectives of record and replay include (1) low overhead during recording, (2) interactive operation during replay, (3) low disk usage, and (4) sufficiently robust technology. Low overhead indicates that the recording process cannot affect the operation of the ISC or consume too many computing resources. Interactive operations are required when replaying the recorded air situation and operations of controllers. It is meaningful for accident analysis. For example, users can measure the distance between two airplanes when replaying the recorded operations. The recorded files cannot occupy much disk space, as long-term recording will exhaust disk space. The recording and replaying process must be implemented by a reliable and robust technology. Some methods to implement a record and replay system are as follows:

Screen capture: This method captures the screen in a video file. All operations of the controller and air situation are recorded. This method is simple to implement. However, it has several drawbacks: (1) The recording requires too much CPU usage. (2) The recorded video files are too large, so a common PC cannot store recorded files for very long. (3) When replaying the air situation, the controller cannot intervene. For example, the controller may want to measure the distance between two airplanes or the distance from airplane to ground when replaying an air accident.

Windows hook: This method takes advantage of the message-driven mechanism of Windows OS. All the messages processed by the ISC can be hooked and recorded in a file. When replaying the air situation, we can open the file and get the message and its time, then send the message to the ISC. The advantage of this method is obvious. The recorded file is small and the overhead of the recording process is relatively small. However, this is a complex and error-prone method. In principle, the ISC responds to the stored Windows message and reexecutes the command of the controller and shows the recorded track data from radar or ADS-B data. If there is any error in the stored messages or track data, critical failure in the ISC will be inevitable. The disadvantage lies in that this is not a robust way to redisplay an air situation. In addition, we cannot replay the air situation from arbitrary time instances by using this method. Moreover, this is a platform-specific method.

Another method is using the Xwindow protocol of Unix OS. The recording application uses Xwindow protocol to store the state of application in Xserver. This is a platform-specific method like the Windows hook method, and we also cannot replay the air situation from arbitrary times.

Obviously, these schemes cannot meet our objectives. Thus, we do not adopt these methods to implement the record and replay module of the ISC system in this chapter. Instead, we record the original operational data of the ISC, including the following four parts: (1) operation time; (2) operation type, including window-related operations, track-related operations, and track plate-related operations; (3) target UI object, including the IDs of controls and its parent controls; and (4) user's input; this refers to the results applied to the ISC. For example, the input in an edit box control will be recorded. A sample of recorded contents is summarized in Table 3.2.

TABLE 3.2

Recording Sample

Operation of Controller	Recorded Content
Main window operation	IDs of controller
Child window operation	IDs of controller and its parent window
Select/deselect track	Position of track in the linked list, flag bit
Modification of track plate	Position of track in the linked list, modified value
Modification of flight plan	International no., IDs of modified controls
Online drawing	Color, width and line shape and positions of each vertex

We record the operations of the controller in the manner of incremental recording, that is, if the controller does not operate for a period of time, nothing is recorded. This method can significantly save disk space.

We conduct four experiments to test the efficiency of our approach against others in our ATC platform. We specially test the disk usage of each solution, because CPU usage test is only required when there is a large overload, that is, when too many operations need to be recorded. This occurs rarely in general aviation. In our test, a typical air traffic management process is run on our ATC. Two airplanes fly from one place to another in four rounds.

The four rounds cost about 20, 40, 80, and 160 min, respectively. During the flying, the operations of controller are recorded by each method, and disk usage is recorded respectively. The test results are shown in Figures 3.12 and 3.13.

From Figures 3.12 and 3.13, we can see that disk usage in our method is greatly less than others. In addition, the curve in Figure 3.13 shows a flattening trend. This is to say our method utilizes less disk space when the flight period increases. This is due to our incremental recording design as mentioned earlier. This can be interpreted as follows: Any flight will require specific interactions with the controller, such as departure report, flight plan manipulation, landing report, and so on. Usually, the controller will rarely intervene in the flight plan, except in dangerous situations. As the traveling time increases, the idle time of controller will increase. The idle time refers to the period when the controller performs no operations.

Several fault-tolerance techniques are implemented in the record and replay module to improve its robustness. The duration of the recorded files does not exceed 5 min. In this way, not too much information is lost even if faults occur during replaying. Moreover, we record the "global state" of the ISC at the head of each recorded file, including the state of all windows, controls and maps, tracks, etc. The recorded global state can be treated as a

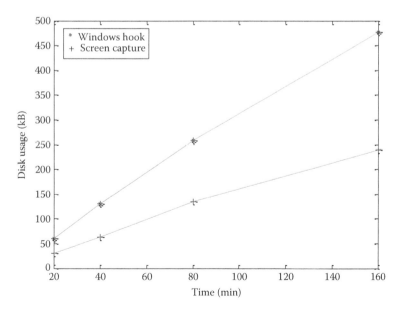

FIGURE 3.12
Disk usage of other solutions.

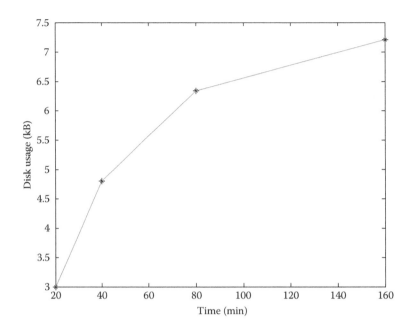

FIGURE 3.13
Disk usage of our solution.

savepoint. If any faults occur during the replaying process, we can roll back to the save-point then retry. These measures can significantly enhance reliability.

Considering the interactive operation during the replay of an air situation, this method allows the controller to intervene in the replaying process. For example, the controller can measure the distance between two airplanes, zoom in/out on the screen, etc. These operations can help the controller to make an in-depth investigation of the cause of air accidents.

3.6.5 Handover

When an airplane flies very long distances, it will be controlled by several air traffic centers until it reaches the destination airport. Hence, the airspace is divided into many controlled zones, each controlled by an air traffic center. When the airplane flies to the boundary of two air control zones, the ATC system will automatically detect this and the track on the ISC will start to flicker. It should be noted that, the primary radars set in both control zones will see the airplane, and the controllers in both air traffic centers will see the flickering tracks. Usually, the next controller will manually accept the airplane by pressing the "ACC" button on the menu pop-up by right-clicking the flickering track. In fact, the handover process is to transit the jurisdiction of the flight plan from the current ATC center to the next one.

3.6.6 Low-Latitude Airspace Infrared Supervision

In low-latitude airspace, general aviation also should be guided. For general aviation, ADS-B is the primary communication technique between controller and pilot. The advantage of

ADS-B is two-fold: (1) it is very economical; and (2) it can provide rich and accurate information about the airplane, such as the speed and operating state of the airplane. Nevertheless, the disadvantage of ADS-B is obvious. Because ADS-B uses a one-way broadcasting technology, the receiver need not respond to the requests from the ADS-B. Hence a hostile flying object carrying ADS-B can disguise itself as a common airplane by sending fake datagrams. For example, if an invasive fighter sends false datagrams to the receiver of the ATC, the controller cannot identify it. It should be noted that the primary radar can only detect but not identify the flying object. More specifically, based on the available information, the primary radar can only get the position and speed of the flying object, but not what it is.

Thus, we need other surveillance techniques to identify unidentified flying objects. Infrared surveillance can show the shape, size, flying altitude, alpha, position, etc. [6]. In sensitive areas such as border area, infrared surveillance can be used as a complementary surveillance technique for the ATC.

Infrared surveillance systems include an infrared camera that is usually mounted with the primary radar. Because an infrared camera cannot endure long-term operation, it is in hot standby state. When the primary radar detects an unidentified object, the infrared camera will start recording automatically [7]. With the positional information from the primary radar, the infrared camera can easily capture the flying object.

The rationale behind the integration of the primary radar and infrared camera lies in the fact that: first, we use primary radar to detect flying objects and infrared camera to identify them; second, infrared camera cannot locate the flying object at startup state without the assistance of the primary radar; and finally, the infrared camera cannot endure long working times—most of the time it is ready but not working. The integration of both will make the detection and identification better.

When the infrared camera finds the flying object, it will capture it on video. The video stream will be transferred to the ISC by network. The dataflow in a typical infrared surveillance is shown in Figure 3.14.

Figure 3.14 illustrates the collection, transmission, and display of an infrared video stream. There are several coding standards for video coding and compression. More specifically, H.263 is usually used for video meeting and MPEG-1 and MPEG-2 for multimedia applications. In this chapter, we use H.263 coding standard due to its real-time property. Finally, the infrared video stream will be received and shown by ISC, with which the controller observes and identifies the flying object by visual means.

The infrared surveillance window of ISC is shown in Figure 3.15, in which a flying airplane is captured by an infrared camera. In Figure 3.15, we can clearly see the shape of the flying object.

It should be pointed out that, Figure 3.15 only shows a screenshot of the video. In fact, the infrared surveillance window can display the full video. As mentioned earlier, when

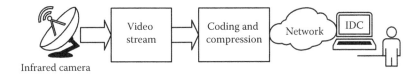

FIGURE 3.14
Infrared video transmission.

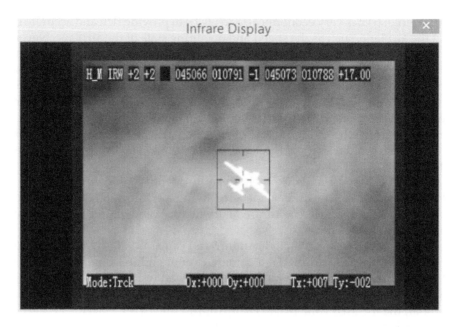

FIGURE 3.15
Infrared display window.

starting, the infrared camera needs the help of the primary radar to locate the suspected flying object. Once the object is located, the infrared camera can automatically track the flying object by means of some image recognition.

3.7 Conclusions/Outlook

In this chapter, we introduce the modern ATC system from the perspectives of the user and developer. The composition of the ATC system is illustrated, and some peripheral equipments such as ADS-B, secondary surveillance radar, and surface movement radar are also mentioned. Some design experiences are presented in this chapter, including the considerations when selecting a better method.

All subsystems of the ATC system are described in the order of data flow. First, we describe how to extract positional information from a radar data package and translate it to track. Then the positional information will be used to calculate various conflicts. Management of the flight plan is illustrated. Finally, surveillance data display and ground-to-air communication are described in detail. In addition, we introduce the design and implementation of the record and replay module, including how to increase the performance, availability, and reliability.

To sum up, the operating mechanism of the ATC system is clearly understood. After reading this chapter, you will know how to design an ATC system and how to use it to control in-flight airplanes.

References

1. Li, R.X., Bar-Shalom, Y. Design of an interacting multiple model algorithm for air traffic control tracking. *IEEE Transaction on Control System Technology*, 1, 186–194, 1993.
2. Dimitris, B., Lulli, G., Odoni, A. An integer optimization approach to large-scale air traffic flow management. *Operation Research*, 59, 211–227, 2011.
3. Zhang, W., Kamgarpour, M., Sun, D., Tomlin, C.J. A hierarchical flight planning framework for air traffic management. *Proceedings of the IEEE*, 100, 179–194, 2012.
4. Everdij, M.H.C., Scholte, J.J., Blom, A.P., Stroeve, S.H. An investigation of emergent behaviour viewpoints in literature, and their usability in Air Traffic Management. 2011. http://reports.nlr.nl:8080/xmlui/bitstream/handle/10921/144/TP-2011-444.pdf?sequence=1. Accessed March 28, 2014.
5. Curtis, S., Hallett, E., Mirchi, T. Training air traffic controllers for future next generation air transportation system (NextGen) technologies. California State University, 2014. http://chaat.cla.csulb.edu/research/Publications/Curtis%20Hallett%20Mirchi.pdf. Accessed March 28, 2014.
6. FAA (Federal Aviation Administration). 91-36D—Visual Flight Rules (VFR) flight near noise-sensitive areas. 2004. http://www.faa.gov/regulations_policies/advisory_circulars/index.cfm/go/document.information/documentid/23156. Accessed March 28, 2014.
7. Li, J., Gong, W. Real time pedestrian tracking using thermal infrared. *Imagery Sensors*, 5(10), 1597–1605, 2010.

4

Security of VHF Data Link in ATM

Meng Yue

CONTENTS

4.1 Introduction

Digital air–ground communication allows transfer of digital information between equipment onboard an aircraft in flight and ground-based computers connected through a data link service provider's (DSP's) transport network. There are many ways to support the data link, such as line-of-sight very high frequency (VHF), high frequency (HF), and satellite. As data link use increases, it becomes necessary to have data link channels constantly available.[1]

The Aircraft Communications Addressing and Reporting System an (ACARS) is a data link system that allows communication of character-oriented data between aircraft systems and ground systems. It enables the aircraft to operate as part of the airline's command, control, and management system.

Equipment required onboard for data link is known as an ACARS management unit (MU). The MU is connected to a standard airborne transceiver for data link communication and may be connected to other airborne equipment via an airborne communication bus. The communications management unit (CMU) provides ACARS communications function in lieu of an ACARS MU. The CMU is designed to support bit-oriented message transfer utilizing the protocols and procedures for routing. The CMU will interact with onboard systems that supply mobile air–ground communications over VHF, satellite, and HF media.[2]

There are two parts in this chapter, the ACARS principle and message format are described first and the ACARS data link message security protection method is detailed in the second part.

4.2 ACARS System Principle and Message Format

ACARS, as an important means of current aeronautical communication system, is widely used in aviation and is essential to ensure flight security. This section will detail the composition of ACARS, how it works, and the ACARS message format.

4.2.1 Overview

ACARS is a character-oriented data link, which cannot transmit digital voice and data stream files, such as satellite cloud images. This system consists of an airborne subsystem (management unit and control unit), VHF remote ground station (RGS), ground data communication network, ground network management and information processing system (NMDPS), and data link users.[3]

Over the years, people always want to provide the latest information to the aircrew through the voice communication system without adding an extra burden. ACARS enables the aircrew to send information to the route command center, such as departure time, arrival time, fuel situation, departure delay, and other information. The ACARS system remains aircrew workload tractable and provides the aircrew and route command center with information services at the same time. Other advantages of ACARS include ground surveillance capabilities on aircraft engine and other parameters, a more effective exchange of information related to flight arrival and sustained flight, reducing multiple frequency changes in aircraft, and a more reliable selective call system.

At present, the main international data link service providers include Aeronautical Radio Incorporated (ARINC) (United States), Societe Internationale de Telecommunications Aeronautique (SITA) (Europe), and Automatic Data Direction Control (ADCC) (China), and they have different data link service area coverage. Each airline can choose a different DSP to provide services. When the aircraft flies through different service areas, it needs to deliver information through the international gateway.

4.2.2 Research on the Composition and Operation of ACARS

This section details the composition of the ACARS data link communications system and describes how ACARS works.

4.2.2.1 Composition of ACARS

ACARS consists of airborne subsystems and ground stations network. The ACARS airborne subsystem is composed of VHF transceiver, VHF, VHF antenna, control unit, MU and ACARS control unit (CU), as shown in Figure 4.1. The ground station network is composed of VHF ground station, a central processing computer and a conversion network connected with the computers of each route.

ACARS is a collaborative system providing air–ground and ground–air digital voice and data communications. The ACARS network provides a common data link that can handle all kinds of messages. The ACARS control management unit enables airborne systems transmit and receive work.[4]

ACARS is an air–ground communication network, which can connect aircraft as a mobile terminal with the route command, control, and management system. Information about the departure and arrival time, departure delayed information, fuel conditions, and other flight-related data is automatically collected and sent to the system through the control unit. The system uses International Organization for Standardization (ISO) letter, fifth character, and this format is fully compatible with the agreement of the United States Air Transport Association (ATA) and the International Air Transport Association (IATA).

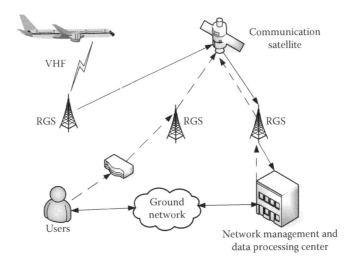

FIGURE 4.1
(See color insert.) The composition of ACARS.

Maximum block length is 220 characters, and a message longer than 220 characters will be divided into a number of blocks.

The message format consists of seven units in accordance with the International Civil Aviation Organization (ICAO) conventions of encoded characters. Eighth is coupled with a character to provide parity bit.

Information is sent from the plane transmitted via 131,550 MHz ACARS data link to terrestrial radio stations, and then it is relayed to a central computer processor where the information is converted into a route-task message. The route-task message is sent to the relevant routes by ARINC electronic switching system (ESS).

4.2.2.2 Operation Mode of ACARS

ACARS has two work modes: request mode and polling mode.[5] Request mode allows airborne- or ground-processor platform to initiate communication. When there appears a scheduled event or a ready command input by the pilot, airborne systems will be automatically ready to launch. Then, after determining the ACARS channel without traffic, the management unit will launch the news. If there is traffic on the ACARS channel, the system will wait before transmitting until the message traffic is eliminated. If two or more airborne systems choose the same moment to transmit, the messages will be chaos. In order to avoid transmission synchronization, the messages in the future will be reemitted at random intervals.

As soon as the ground station processor receives a transmission message, it completes the error test of block check sequence (BCS). If the message is errorless, the processor will send it to its proper destination. Ground station processor also sends a confirmation signal <ACK>, which can notice the airborne to eliminate the original message and return to the normal work. If the message is wrong, the processor does not send the confirmation signal and airborne systems will reemit messages. If the message has not been confirmed after six times, the system will warn the aircrew. If the uplink transmission (ground-to-air) is wrong, the downlink message sends a negative signal <NAK>.

The polling mode only allows the ground processor to initiate communication. During the polling mode, airborne ACARS can only respond to the received uplink message (polling). Once the airborne system is in the polling mode, the ground processor maintains a continuous communication with airborne system by transmitting a general polling. When the radio channel is unblocked, the airborne system will respond to these polling, or, if no message, it will reply with the general response of the polling mode. BCS error detection is performed in all messages (uplink and downlink) and generates a confirmation or a denial message (ACK or NAK) included in the next transmission. Ground processor sends a general response of request to command the airborne systems switch to the request mode. When ON event (landing) appears, airborne system work in polling mode will automatically enter the request mode.

For all sampling data during the course of scheduled flights, MU has played the role of organizer or format converter. The management unit collects the data from control units, aircraft sensors, and out, off, on, and in (OOOI) events sensor. The OOOI event sensors determine the flight arrival, departure, and flight time. Management unit generates a Greenwich Mean Time clock to record the occurrence time of OOOI events. At the request of the ground processor, the flight information, cumulated by cockpit flight management computer (FMC), integrated data systems, and aircraft flight recorder terminal, is formatted as a digitally encoded signal and is transmitted to ground processor via VHF transceiver machine.

All uplink (ground–air) emissions are monitored by the management unit to determine the address. The message's destination address is identified through the aircraft registration number or flight number. Aircraft registration number is identified by the unique lapping form of aircraft wiring harness. Flight number is entered into the memory management unit by the flight crew before the flight. Unless the address of registration number or flight number matches the aircraft name, the management unit will not respond to any uplink message.

ACARS uses 131 and 550 MHz horizon VHF radio frequencies according to the current provisions. Additional frequencies can be added as the demands increase. ACARS in HF frequency range is in favor of the extension of air digital services when aircrafts fly over the ocean.

4.2.2.3 How the Airborne Management Unit Works

The management unit is the processor for receiving data from a number of sources. They must also be easy for future expansion when data link system functions increase. So, the management unit is basically a central computer or processor with a variety of other data processor interfaces. These data processors are subordinate to the main processor and do not conflict with the unit, so that the main processing function can compile all the flight data and format them.

The main processor uses the bus arbitration logic to communicate with the slave processor. This kind of signal exchange control form makes main processors read data from the secondary processor and send data to the main processor memory. In addition, the main processor also generates a continuous Greenwich Mean Time for internal and external clock and communicates with the control unit. Monitor and keyboard information is received via the serial input bus and stored in the data buffer. Once the data buffer is full, the main processor will do the error checking of the display information. In order to control the display of the control unit, the main processor compiles the data and transmits it through the serial bus to the control unit. Display information containing specified position for fault management unit or control unit advertised.

The input–output processor monitors OOOI event input and accesses overlapping information from the aircraft wiring system. This information includes the aircraft registration number, route identification, OOOI sensor leads, and program management unit program leader. The input–output processor is also used to send Greenwich Mean Time information to the control unit for display or to the serial output bus for external use in data link system.

The modulation–demodulation processor is used as the interface between data link system and VHF communication system. The modulation–demodulation processor is composed of a demodulator, and the processor must check whether parity uplink message and address (valid flight number or aircraft identification) are correct, and whether the BCS code exists.

4.2.2.4 How the Ground Network Works

A downlink transmission from an onboard ACARS unit is demodulated at the radio terminal and stored in the local memory buffer. This downlink transmission asks each radio terminal buffer at the central processor AFEFS (ACARS front end processing system) whether they concentrate the received message. Message structure is the same as the message structure of the radio link. The message format between AFEFS and ESS adopts the agreement of ATA and IATA. Each message consists of the necessary addressing information, routing information, and free text (if used). The message structure is arranged by the standard message identifier (SMI) and the body of the identifier (TEI). AFEFS completes all the functions related to link control, message formats, message queuing, and network monitoring. AFEFS both receives and transmits messages from the ESS.

4.2.3 ACARS Message Format

Data link communication uses the air–ground network to transmit a certain telegram meeting format in the space. The ground–air communication format should conform to ARINC Specification 618 and the ground–ground communication format should conform to ARINC Specification 620.

4.2.3.1 ARINC Specification 618 and the ACARS Air–Ground Message Format

- The main content of ARINC Specification 618
- ARINC Specification 618, also known as air–ground character-oriented protocol specification, includes ACARS system description, air–ground block structure and message handling protocols, link management for a VHF air–ground network, the air–ground satellite link protocol, and HF data protocol for transporting character-oriented ACARS messages. The ACARS message is introduced according to the definition of the ISO 5 character set.[6]
- The air–ground block format.
- The format of the air–ground message is governed by ARINC Specification 618.[1] The downlink block format is as shown in Table 4.1. The uplink block format is as shown in Table 4.2. The message consists of header, text, suffix, and BCS. The header includes start of header (SOH), mode, address, technical acknowledgment, label, downlink block identifier/uplink block identifier (DBI/UBI), and STX (start of text).

The text field of all messages must consist of the noncontrol characters of the ISO 5 character set. The maximum length is 220 characters. Longer messages, known as multiblock messages, will be divided into several separate ACARS messages to send.

For a single block message, its trailer is with the end of the text (ETX). For multiblock messages, the trailers of all blocks are with the control character end of block (ETB), except that the trailer of the final block is with the ETX.

The BCS is the result of a cyclic redundancy check (CRC) computation to ensure that the block is error free. The BCS is initiated by, but does not include, the SOH character, and is terminated by, and does include, the ETB or ETX character. The BCS is generated on the entire message, including parity bits.[7]

1. SOH: 1 character

   This character is represented by the control character SOH in ISO 5 character set. An SOH character (ASCII code value "0×01") indicates the start of the message header. SOH is also used to indicate the beginning of the text. This text is evaluated by BCS but should not be checked.

2. Mode character: 1 character

   The Mode characters are divided into two basic categories as follows:

   Category A—This category is denoted by a "2" character. Aircraft may broadcast the message to all DSP ground stations in the aircraft's VHF coverage area.

TABLE 4.1

ARINC618: The General Format of Air–Ground Downlink Messages

Name	SOH	Mode	Address	TAK	Label	DBI	STX	MSN	Flight ID	ApplText	Suffix	BCS	BCSSuffix
Size	1	2	7	1	2	1	1	4	6	0~210	1	2	1
Example	<SOH>	2	.N123XX		5Z	2	<STX>	M01A	XX0000	DOWNLINK	<ETX>		<DEL>

TABLE 4.2

ARINC618: The General Format of Ground–Air Uplink Messages

Name	SOH	Mode	Address	TAK	Label	UBI	STX	ApplText	Suffix	BCS	BCSSuffix
Size	1	2	7	1	2	1	1	0~220	1	2	1
Example	<SOH>	2	.N123XX		10	A	<STX>	UPLINK	<ETX>		<DEL>

Category B—Aircraft may transmit the message to a single DSP ground station in the aircraft's VHF coverage area. The address of ground station is placed in Mode character position and is represented by <@> or <]>. The actual Mode character inserted in each downlink message is determined by the ground system access code with a prefix. This access code is selected by MU/CMU based on the uplink Mode character.

3. Address: 7 characters

It is the aircraft registration mark (tail number). MU/CMU should not transmit any downlink message having a valid aircraft registration mark. Once the CMU acquires the aircraft registration mark, it should record it until power to the CMU is removed. The address can identify the aircraft communicating with the aircraft processor. In air–ground messages, the Address field should contain the aircraft registration mark. The aircraft registration mark may be provided by wiring at the MU's interface within the aircraft (typical of characteristic 597 and 724 MU's) or it may be provided in the form of ARINC 429 data words by an external subsystem (typical of the ARINC 724B MU).

4. Acknowledgment: 1 character

The positive technical acknowledgment character consists of an A–Z, or a–z, or a NAK control character. It is the acknowledgement for the message. The sender needs an acknowledgement from the recipient after sending a message. ANRINC618 provides a dedicated acknowledgment reply message to confirm a message that has been received. The technical acknowledgment message contains Label characters DEL. If the recipient needs the feedback of some types of packets from the sender after receiving the message, it can set TAK field to < ACK >, showing that the receiver message is received successfully. When the sender learns that the previously sent message was successfully received, it needs to confirm the message with <ACK> from the recipient, hence reciprocating, until one of them uses <_DEL>. Generally, the TAK character in the first message of a complete communication is <NAK>, and the following is <_DEL> message or a message containing some information with TAK character <ACK>. If it is <_DEL>, the TAK field of the received message will be <BI> field.

5. Label: 2 characters

This Label is in the message preamble. The Label character is used as a shorthand method to describe the message. The Label field indicates the classification of the content of the message being transmitted, and it is used to determine routing and addressing. A list of assigned labels is defined in ARINC specification 620, such as Out/Return IN Report (IATA Airport Code)—Label QG, and IN/Fuel Report (IATA Airport Code)—Label QD. The corresponding meaning and the format of different categories of message are not the same.

6. Downlink block identifier/uplink block identifier (DBI/UBI): 1 character

ARINC618 agreement provides that the <BI> field of the current message being sent cannot be the same as the <BI> field of the previous message. The uplink block identifier (UBI) should consist of an "A"–"Z" or "a"–"z" or single NUL character. The downlink block identifier (DBI) character should consist of a 0–9 character.

7. STX: 1 character

   The control character STX indicates the end of preamble and the beginning of the text.

8. Text: Not exceeding 220 characters in length

   The Text field of all messages must consist of the noncontrol characters of the ISO 5 character set. The maximum length is 220 characters. Longer messages, known as multiblock messages will be divided into several separate ACARS messages to send. A message sequence number (MSN) and flight identifier (FI) must be included in the downlink message. The MSN is a four-character field used for ground-based message reassembly. The FI consists of a two-character Airline Identifier and a four-character Flight Number field. MU/CMU should not transmit any downlink message having a valid aircraft identification. There are sublabels and addresses included in some downlink messages used to address.

9. Suffix: 1 character

   Each single block and the last block of multiblock should be terminated with the control character ETX. The other blocks of multiblock should be terminated with the control character ETB.

10. BCS: 16 bits

    The BCS is the result of a CRC computation to ensure that the block is error free. The BCS is used for the characters from SOH character to ETX or ETB character, but does not include SOH character.

11. Control character DEL: 1 character

DEL is BCS suffix, transmitted following the BCS. The purpose of DEL is to enable the last bit of the BCS to be decoded.

Conclusions from the segment definition are as follows: logic routing and addressing functions implemented in the message, messages in the downlink realizing restructuring on the ground, ensuring the accuracy, timeliness and completeness of the message, and laying a good foundation for the NMDPS to process and distribute message.

4.2.3.2 ARINC Specification 620 and the ACARS Ground–Ground Message Format

1. The main content of ARINC Specification 620

   Data link users, aircraft and ground users, use different protocol with DSP. DSP provides not only the appropriate routing information, but also the message format conversion between the two kinds of protocol. ARINC620 describes the air–ground communication protocol and ground–air communication protocol, the function input and output of DSP systems from the data link user's perspective.[8]

   ARINC620 agreement mainly involves DSP functions, such as information transfer, the flight tracking, network management, and DSP access.

2. The ground–ground message format

   The message processed by DSP can be divided into two categories according to the flow of information on the data link: uplink message, from the ground user to the aircraft, and downlink information, from the aircraft to ground user.

DSP processing information relates to two aspects. DSP converts the downlink message sent by the aircraft in accordance with ground–ground message format, and then sends it to the ground user via ground communication network. DSP converts the uplink message sent by the ground user in accordance with air–ground message format, and then sends it to the aircraft via data link. Characters used in ACARS messages are limited to ISO 5 character set.[9]

1. Downlink ground–ground message format

   The DSP receives the downlink messages from the airborne through a ground station. The downlink message format conforms to ARINC Specification 618. The DSP converts the received ACARS message according to ground–ground message format in the ARINC Specification 618, and then delivers the message to the user via ground communication network.

   When the DSP receives a downlink ACARS message from the ground station, the Label within the ACARS message indicates whether the DSP generates a ground–ground message and sends it to the ground user. The general format of ground–ground downlink messages is shown as Table 4.3.[8]

   Line 1, the Priority/Destination Address line, also known as simply the Address line. The contents of the line consist of two parts: the characters of the priority of the ground message and the address list of the information recipient. Each message priority identifier consists of two characters, followed by a *space* and then the destination address list including seven characters. The maximum number of addresses is 16.

   Line 2, the address and sending time of the sender. This line begins with the *period* character <.> and is followed by the address and a timestamp in the format day/hour/minute (ddhhmm). It is possible to enter further signature information after the timestamp.

   Line 3, the SMI line. It contains a three-character code and it is a part of message type flags.

   Line 4, the Text Element field of the message. It is a series of text elements. Each text element is composed of three parts: text element identifier (TEI), data, and a text element terminator (TET), as shown in Table 4.4.[8] The first text element is usually the FI, which is composed of a two-character airline identifier and a four-character flight number. DSP has the responsibility to translate the three-letter code (as defined by ICAO) airline identifier to the appropriate two-character IATA equivalent.

TABLE 4.3

General Format of Ground–Ground Downlink Messages

Line	Contents	Example
1	Priority/destination address	QU ADRDPAL
2	Signature/transmission time	DSPXXXX 121212
3	SMI	AGM
4	Text elements	FI XX0001/AN N123XX
5	Communication service line	DT DSPRGS121212M01A
6-n	Free text	- DOWNLINK

TABLE 4.4

Text Element Structure

Field	Length	Coding
Text element identifier (TEI)	2 characters	Alpha/numeric
Text element delimiter	1 character	Space
Text element data field	Variable (depends on TEI)	
Text element terminator	1 character	</> If another text element follows or <cr/lf> if this is the final text element

Line 5, the communication service line, contains the text element identifier DT and the final block. It consists of four fields:

a. DSP identifier. The identity of the DSP provider can be got from this field.

b. If the ground station receiving ACARS downlink messages uses ARINC618 for data transmission, it means this station have received the first block of the downlink message.

c. Message reception time stamp.

d. Message sequence number as provided in the downlink ACARS message.

Line 6, free text. It is optional and is not part of a message's structured text.

2. Uplink ground–ground message format

DSP converts the uplink message sent by ground user to the ACARS uplink block, and sends it to the specified aircraft via ground station.

Uplink ground–ground messages received by the DSP for transmission to an aircraft are accepted if the following conditions are fulfilled:

a. The timestamp of the message indicates the effective time.

b. The SMI is valid and is approved by the ACARS user.

c. The SMT contains either an aircraft registration number (AN) text element or a FI text element and the corresponding valid text. If the DSP does not have tracking information for the aircraft addressed, the following supplementary condition applies:

 i. If the SMT contains either a GL text element (approximate geographic location of aircraft) or an AP text element (airport location of aircraft) and an airport or city that DSP can identify, the DSP will determine the ground station for transmission to the aircraft based on this information.

 ii. The application text is preceded by the special TEI—also referred to as dash space <-sp>, which is used as a separation.

 iii. The Line 1 of the ground–ground message must contain the address of the DSP.

If the ground uplink message sent by the user cannot meet the above conditions, DSP will intercept the information and send it coupled with the reason of intercepting. The general format of the ground–ground uplink message is shown in Table 4.5.[8]

The structure of uplink messages in Table 4.5 is the same with the structure of uplink messages in Table 4.3. Table 4.4 describes the Line 4 (text elements) in the Table 4.3.

TABLE 4.5

General Format of Ground–Ground Uplink Messages

Line	Contents	Example
1	Priority/destination address	QU CTYDPAL
2	Signature/transmission time	.QXSXMXS
3	SMI	AGM
4	Text elements	AN N123/MA 123A
5-n	Free text	- UPLINK

DSP can get the address of the aircraft in the uplink message contained in aircraft tail number (AN) and FI. When text includes transmission path (TP), it indicates that the user designates a transmission path for the uplink message. When the aircraft is not in flight, DSP can send the message to the specified aircraft via ground station or airport that can be gotten from a geographic locator (GL) text element or an airport locator (AP) text element. When the uplink message includes message assurance (MA), it means that the user requires the confirmation after DSP delivers the message to the aircraft.

4.2.4 Conclusion

This section introduces the working principle and the structure of the ACARS data link system and makes a detailed analysis about air–ground communication format (ARINC618) and ground–ground communication format (ARINC620) in the ACARS data link.

4.3 ACARS Datalink Message Security Protection Method

Without suitable protection, ACARS datalink communications are vulnerable to threats posed by unauthorized entities that may access or modify message content, the result of which may be to expose sensitive information or endanger the safety and integrity of aircraft operations. This chapter will make a detailed analysis of the vulnerabilities of the ACARS datalink. Also according to the characteristics of the ACARS datalink, the practical ACARS datalink message security protection method will be proposed.

4.3.1 Security Threats of ACARS Datalink

The ACARS datalink belongs to the wireless network and compared with wired networks, it is more vulnerable. All ACARS messages are completely transferred in the form of plaintext, and anyone can get almost all the ACARS information very conveniently using a computer, sound card, or radio frequency (RF) antenna and freely available software. Also the existing software can simulate aircraft or controller terminals to allow calls to the real terminals. China does not attach importance to the security of ACARS system. There are no any safety precautions in the air-to-ground VHF link to protect the safety of the link, but only with a simple user management in the user terminal. This method can only protect the system from unintentional illegal intrusion. When faced with a malicious attack, the system will be helpless. So it cannot effectively ensure the security of the system. In a nutshell, without suitable protection, there are currently a great number of security threats to the ACARS network.[10]

4.3.1.1 Data Leakage

Existing ACARS datalink security is so poor that any attacker who has an RF transceiver can intercept ACARS message and understand its contents. ACARS data information of civil aviation contains a series of airlines-sensitive information, such as take-off and landing data, oil data, and unit data. If an attacker intercepted and analyzed the data, the interests of the airline may be directly affected and cause huge economic losses. At present, more than 70% passenger aircraft in China are equipped with ACARS, and airlines are more and more dependent on it to ensure flight safety and significantly improve the company's operating efficiency. At the same time, higher demands for the safe and stable operation of ACARS are put forward. According to the regulations of the Civil Aviation Administration of China, all aircraft should be equipped with ACARS airborne electronic devices and applications of ACARS will continuously expand as military airplanes are also asked to add ACARS airborne communications systems. As military aircraft have high confidentiality, the consequences of data leakage would be unimaginable. Currently, ACARS messages sent by planes can be intercepted by using ACARS receiving software and a simple wireless receiving device. There are common software tools, such as WACARS and acarsd, which is shown in Figure 4.2.

Data leakage may not directly affect flight safety, but it has become a great hidden trouble for ACARS datalink.[11]

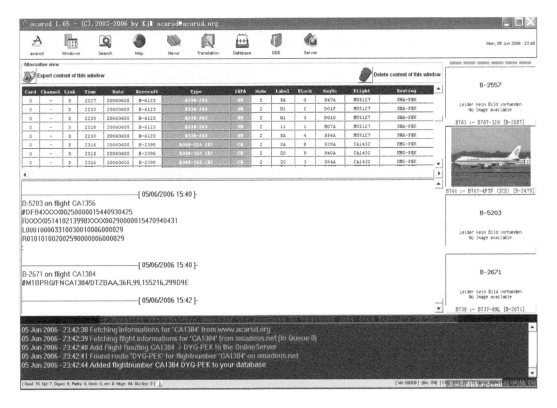

FIGURE 4.2
(See color insert.) ACARS receiving software acarsd 1.65.

4.3.1.2 Data Deceiving

Because of the lack of effective security protection, spurious ACARS information could be transmitted through the ACARS datalink and can appear to be legitimate. Even effective ACARS information is also very likely to be tampered or resent during transmission, leading to data error and directly impact on aircraft safety. Now, ACARS messages contain CRC check mechanism that only allows code error testing in the process of transmission and does not have the ability to detect man-made deliberate manipulation. Because it is impossible to confirm message integrity, it cannot be determined whether the message received is actually from the legitimate sender. Data deceiving will directly affect flight safety, as the message recipient cannot identify the authenticity of the information.

4.3.1.3 Entity Masquerade

In ACARS, an entity can easily be disguised as a terminal to damage the air-to-ground communication and to hinder the normal operation of the system. For example, a computer with simple equipment can simulate a controller to send illegal control messages to the aircraft and it is very easy to cause a catastrophe such as collision, etc.

Entity masquerade can be of two kinds as follows:

1. Masquerade as an aircraft

   At present, communications between aircraft and ground receiving equipment need the aircraft tail number and the service code of airlines, as the basis of DSP network operation control center to identify the aircraft. Both pieces of information can be obtained publicly, and then they can be used to masquerade as an airplane through relevant electronic equipment.

2. Masquerade as an airline

   Currently, the ways that link network provides services to airlines are: first, to validate the router interface's IP address of airlines, and then, the system provides further data services only for the legitimate IP addresses (network layer). Second, to validate the username of airlines login (application layer); only through the verification, DSP network control center can finish identity authentication for users, and then provide users with the corresponding data services. It could be found that the two security measures are implemented in different network layers.

In the DSP network operation control center, firewall technology has been widely adopted to strengthen the protection of user data in a datalink network. The DSP network operation control center only allows specified users to communicate by using the specified name, specified communication port, and does not open up the other unspecified rights to users.

Some means cannot fundamentally solve the problem of entity masquerade at present, and entity masquerade is a serious threat to flight safety and a major hidden trouble of ACARS datalink.

4.3.1.4 Denial of Service

There are two methods of denial of service attacks in the ACARS system. In one way as the ground station can only service to one terminal, the attacker can send a lot of false information to the ground station. As a result, the ground station will fail to respond normally to aircraft communications, bringing about a denial of service to the ground station. The other way is that the attacker disguised as a terminal sends lots of useless ACARS information to the datalink, overloading the information processing center server and depleting resources, thus resulting in the server's denial of service. Denial of service attacks can cause the disruption of normal communication, becoming a serious threat to the safety of flight.

4.3.2 ACARS Message Security

Aiming at aforementioned security threats of the ACARS system, ACARS message security (AMS)[2] needs to provide some security services. AMS protects ACARS datalink messages from unauthorized disclosure and modification, and it provides communicating entities with assurance that the source of messages is as claimed.

The three main services are as follows:

1. Data confidentiality

   Data encryption is the most commonly used method to prevent data leakage. The AMS data confidentiality service provides users with assurance that the content of an AMS message is protected from disclosure to an unauthorized third party during data transmission. In end-to-end mode, in order to guarantee correct data delivery, its routing information cannot be encrypted, and only the information body can be encrypted. In addition, identity authentication can also prevent illegal intrusion to the system and avoid data leakage.

2. Data integrity and message authentication

   Data integrity and message authentication services are treated as a composite security service. The authentication mechanisms (e.g., digital signatures and message authentication codes) that make a recipient able to corroborate the data source also provide assurance that data was not modified in transmission. In other words, if the data is authentic, then it cannot have been modified and if the data have been modified, then it cannot be authentic.

   a. Data integrity

      The AMS data integrity service provides communicating entities with assurance that an AMS message is protected against modification after the creation and transmission by an authorized source. Modification includes insertion, substitution, or deletion of message content, either accidental (such as errors caused by a noisy transmission channel) or intentional (such as errors introduced by an unauthorized entity).

   b. Message authentication

      The AMS message authentication is also known as data origin authentication, and it can effectively mitigate the spoofing threats to provide a message recipient with assurance that the source of an AMS message is as-claimed. In other words, message authentication provides the aircraft crew or an automated aircraft information system with confidence that the messages they received originate from the claimed ground source, for example, DSP or airline operations. Similarly, it will provide a dispatcher or automated ground information system with confidence that messages they received originate from the claimed airborne source such as an aircraft information system.

Data integrity and message authentication security services can ensure that the service objects are authentic and effective terminals, and prevent junk information transmitted in the datalink, so it can also effectively prevent denial of service attacks.

3. Key establishment

The AMS key establishment service provides communicating entities with the ability to establish cryptographic keys in a secure and authenticated way. The keys are necessary to support the data confidentiality, data integrity, and message authentication services.[12]

4.3.3 ACARS Information Security Architecture

ACARS information security architecture contains two modes: air–ground security architecture based on DSP and end-to-end security architecture.

4.3.3.1 DSP-Based Security Architecture

The DSP-based security architecture provides security assurance by DSP, relying on the ACARS ground host and the PKI/CA to distribute the keys. The system architecture provides a security envelope between an aircraft avionics system and a DSP ground system that implements AMS, as shown in Figure 4.3.[2] The AMS-protected message is protected during the transmission in the air–ground subnetwork. When the AMS-protected message is processed, the original message is recovered, and will be formatted and routed by DSP. The formatting and routing functions of DSP only apply to existing, nonprotected ACARS messages. So, the DSP may employ non-AMS security services to protect message exchanges when they traverse the ground–ground network between the DSP and the ground systems with which the aircraft entity is exchanging messages.

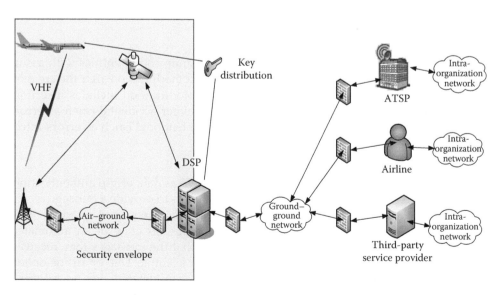

FIGURE 4.3
(See color insert.) DSP-based security architecture.

It is suitable to use DSP-based security for protection of ACARS messages:

- The DSP must be able to read and process ACARS messages to ensure proper operation of the air–ground network.
- The DSP provides the formatting and routing of ACARS messages.
- The destination of ACARS message does not implement AMS.

The demand for DSPs is higher in the DSP-based security architecture. They need to provide not only datalink service to the end users such as airplane, airlines, air traffic control, and so on, but also undertake the responsibility of PKI (public key infrastructure) at the same time, being responsible for key distribution and management, providing encryption, and authentication security services. In this security mode, DSP distributes the initialization key (in the case of using the shared secret key) or digital certificate (in the case of using public/private key). In communication, a MAC is generated by using a shared secret key between DSP and aircraft or making digital signatures using an elliptic curve algorithm to initialize security connection, to complete the basic information exchange and the establishment of a security connection, and to generate the session keys KENC and KMAC, respectively for ACARS message encryption and generation of MAC. Key exchange uses the ECDH algorithm (the elliptic curve Diffie–Hellman), and the session keys are generated by using key generation function KDF. When the connection is established, the two communication sides verify the MAC (generated by using hash message authentication code [HMAC] calculation) first, and then progress or response message after the legal status is mutually determined. In the DSP-based security architecture, encryption and authentication devices are installed on the aircraft and DSP respectively. The security envelope is shown in Figure 4.3.

The advantage of this security mode is that safe operation is transparent for operators (including airlines and the military); it is DSP that centralizes the routing of ACARS information; hence, it will be very difficult for the attacker to get routing information in the air–ground network to make data analysis.

But its drawbacks are also obvious: first of all, DSP must carry out a heavy upgrade to provide security services. Second, there is no internal information security protection in the ground–ground network, so an additional mechanism is needed. Furthermore, data are not secret for DSP, and the plaintexts saved by DSP also become a huge potential safety hazard, and internal attacks within DSP cannot be prevented.

4.3.3.2 *End-to-End Security Architecture*

In the end-to-end security architecture, key distribution and management are carried out by the operating mechanism to ensure the safety of ACARS. The end-to-end security architecture, as shown in Figure 4.4,[2] provides a security envelope between an aircraft avionics system AMS and an airline or third-party service provider ground system. Both terminals implement AMS. AMS-protected message exchanges are protected during the transmission, so the DSP cannot read or process the encrypted content. DSP simply routes the messages to the intended ground entity according to the information taken from the protected message label, airline code, and the airline-specified destination. In this security architecture, the ground system is also responsible for handling ACARS supplementary addresses contained in the protected message and for ACARS message formatting performed by a DSP.

It is suitable to use end-to-end security architecture for the protection of ACARS messages containing airline-proprietary or sensitive information, such as information linked to individuals.

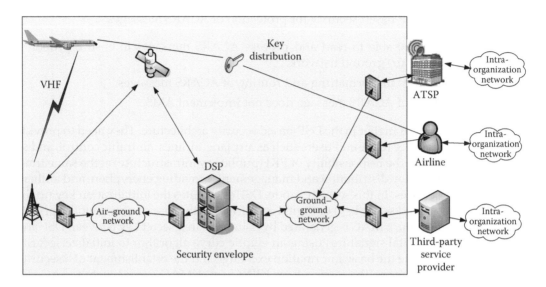

FIGURE 4.4
(See color insert.) End-to-end security architecture.

However, it is not suitable for the protection of ACARS messages that a DSP must be able to read and process to ensure proper operation of the air–ground network.

The main difference between the end-to-end and DSP-based security models is that the former requires the operating mechanism, namely, the airlines, and other end users take the responsibility of key management and distribution. Implementation means on the technical side are the same.

The advantage of this model is to provide end-to-end security, to guarantee the security of the whole link from the aircraft to the ground host of the operator, namely, the air–ground link + ground–ground link. Security protection for each session has realized multiple secure sessions with multiple destinations. Encryption is transparent for DSP and preservation of the ciphertext information by DSP can effectively avoid DSP internal attacks. However, the operator needs to make large changes to protect safety in this way.

The end-to-end model can achieve a higher level of security. However, security protection is needed for each operator itself, which limits its feasibility.

It can be seen that the task of key management and distribution is not taken by professional security agencies in both DSP-based security architecture and end-to-end security architecture and there will be problems on safety and feasibility. As an improvement to the two modes, the task of key management and key distribution can be separated and handed over to a third party, forming a security framework based on the third party. In this way, DSP and terminal users can avoid upgrading for a large security framework. And special security agencies are responsible for key distribution and management that ensure the feasibility of the solution.

4.3.4 AMS Implementation Scheme and Key Technology

AMS needs to provide the following services to protect the security of ACARS system: link encryption service, message authentication service, integrity service, and key management service. And each service can use a different techniques. Link encryption service can use a data encryption algorithm; information authentication and integrity service can use digital signatures or message authentication codes, and the key management service can use a key exchange algorithm and key generation function. However, because the resources of ACARS airborne equipment and the bandwidth of VHF air–ground data link are limited, the appropriate technology should be chosen to achieve the best effect in specific implementation.

4.3.4.1 Data Encryption

In general, encryption technology is divided into symmetric and asymmetric encryption. Symmetric encryption is also called the private key encryption, using the same key to encrypt and decrypt. Asymmetric encryption is also called public-key encryption with two separate but related keys, one of which must be confidential as with symmetric encryption while the other key can be publicly distributed. A comparison of the two kinds of encryption technology is shown in Table 4.6.

Both encryption techniques can be used to encrypt ACARS datalink. Compared with public key encryption technology, due to the advantages of low computational cost and high encryption speed, symmetric encryption technology is more suitable for ACARS, whose resources and bandwidth are limited. The secret key of symmetric encryption must be transmitted in a safe manner, and the period key is short, but it can be solved by the application of key management where the session key is one-time pads.[13]

In addition, ACARS information is character-oriented, but encryption is generally bit-oriented. As a result, the ciphertexts that are nonprintable characters probably cannot be transmitted through the ACARS datalink. This will be introduced in the following coding section.

CFB (cipher feedback) mode in AES (advanced encryption standard) is adopted for the reason that the resource and bandwidth of ACARS system are limited. The length of the key is 128-bit, marked as AES128-CFB128.[14] The main reason considered is that the AES algorithm possesses high safety and CFB mode allows encrypted data with arbitrary length, convenience, and flexibility for use.

AES block cipher receives a 128-bit plaintext, and then creates a 128-bit ciphertext under the control of the 128-bit key. It is an alternative—exchange network design with 16 rounds (Round) iterative process and a plaintext is changed to a ciphertext after the process.

A round of AES is composed of the following four steps:

1. SubByte
2. ShiftRows
3. MixColumns
4. AddRoundKey

TABLE 4.6

Comparison between Symmetric Encryption and Asymmetric Encryption

	Asymmetric Encryption	**Symmetric Encryption**
Performance	The algorithm is complex and the encryption speed is low.	The algorithm is simple and effective with high encryption speed
Key management	Only private key is confidential	Secret key must be established and protected confidentially by both sides
	The confidentiality period of public/private key is long	The confidentiality period of secret key is very short

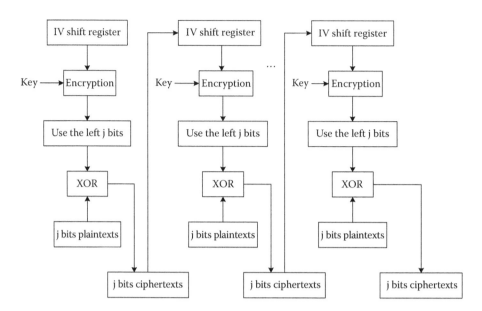

FIGURE 4.5
CFB mode flow diagram.

The 128-bit round key used in each round is generated by the secret key through the process that is referred to as a key schedule.

CFB mode allows the length of encrypted data arbitrary, so it is suitable for character-oriented applications. Every j bits are processed in CFB mode and specific encryption processes are only needed to repeat the following steps, until all plaintext units are encrypted[15]:

The first step is to use the initialization vector IV and key to create encryption vector.

The second step is to make XOR (exclusive or operation) between j bits plaintexts and encrypted vector to get the j bits ciphertexts.

The third step is to move the initialization vector j bits to the left, and to fill j bits ciphertext.

CFB128 is adopted dealing with 128-bit at one time and the specific process is shown in Figure 4.5.

4.3.4.2 Message Authentication Code

MAC is very important in ACARS information security policy because of both the data integrity and message authentication services, but also because the defense of DoS cannot succeed without it.[16]

The purpose of MAC is to ensure that both of the two sides sharing a secret key (or more) in communication have the ability to check whether the transmitted information is modified or not. Further, it can also guarantee the identity authentication between two or more sides in communication.

The news received and secret key are the input of the MAC algorithm that produces a MAC tag with fixed size to finish this work. News and the tag will be transmitted to the receiver, to recalculate the tag and then to compare the transmitted tag with the recalculated tag. If they are same, the news is almost certainly authentic and correct. Otherwise, the news is not right, and should be discarded, or the connection needs to be given up, because it may have been tampered with. The attacker who wants to forge or tamper with the message needs to break the MAC function and it is obviously not an easy thing.

HMAC is selected as the MAC algorithm in ACARS security scheme. HMAC, uses one-way hash function on cryptography and transforms it to MAC algorithm.

It needs a hash function used to encrypt[17] (marked as H, may be MD5 or SHA-1 or SHA-256 is selected in this chapter) and a key to define the HMAC. We use B to represent the number of data block bytes (the word length of segmentation data block of the given hash functions B=64), and use L to represent the number of output data bytes of hash function (L=16 in MD5, L=20 in SHA-1, and L=32 in SHA-256). The length of the authentication key can be any positive integer that is not bigger than the word length of the data block. If the length of the key is bigger than B, the key would be processsed by hash function H. The output of hash function H is a L-length string which can be used in the HMAC. In general, the recommended minimum key length K is L bytes. In general, the recommended minimum key length K is L bytes.

We define two fixed but different strings, ipad and opad: (*i* and *o* represent internal and external)

ipad = the byte 0×36, to repeat B times

opad = the byte $0 \times 5C$, to repeat B times

To calculate HMAC of *text*

HMAC=H(K XOR opad, H(K XOR ipad, text))

As the following steps:

1. To create a B-length string by adding 0 after the key K (for example, if the length of K is 20 bytes, B=64 bytes, then 44 zero bytes 0 x00 should be added after the K)
2. To make XOR between B-length string generated by the first step and ipad
3. To fill the data flow *text* into the result string of step (2)
4. To process the data flow generated in step (3) by H
5. To make XOR between B-length string generated by the first step and opad
6. To fill the result of step (4) into the result of step (5)
7. To process data flow generated in step (6) by H, and the output is the final result

A structure diagram is shown in Figure 4.6.

4.3.4.3 Digital Signature Scheme

The authentication of a message in ACARS requires authentication between peer entities to guarantee reliability. For example, the plane needs authentication from the ground station and the ground station also needs authentication of the information acquired from the plane. The integrity should guarantee the content integrity and sequence integrity of the ACARS message.[18]

By using public key encryption technology, we can proceed with the message authentication and guarantee data integrity. The public key encryption technology includes many applications, such as the RSA algorithm based on the integer factorization problem, the data signature algorithm (DSA) based on the discrete logarithm problem, the Diffie–Hellman

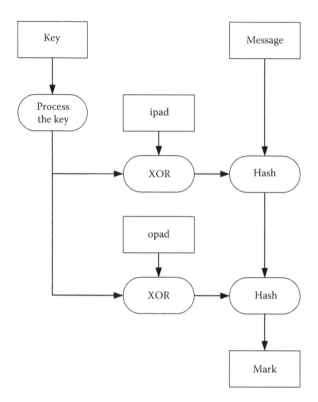

FIGURE 4.6
Structure diagram of HMAC.

(DH) algorithm, the ECDSA algorithm based on the elliptical curve problem, the ECDH algorithm,[19] and so on. RSA algorithm, DSA algorithm and ESDSA algorithm can provide the message authentication service perfectly. The method is that the sender uses the private key to encrypt, and then the receiver uses the public key to decrypt. By this way, as the key is private, the receiver can identify the sender and guarantee the integrity of the message. The security of public key encryption technology depends on the length of the key. For example, the elliptical curve algorithm has a shorter key in the same security rank compared with other algorithms, so it is better adapted to the limited resources of ACARS. Furthermore, the purpose of the design of the data signature is to carry out identity authentication and to avoid deception and denial. Hence, we can choose the ECDSA algorithm.[20]

In the ACARS information security solution, there are two ways to establish a secure connection: shared key and public/private key, and the latter uses the ECDSA as the initial authentication method when it establishes connections. However, in the process of the data transmission, the system mainly uses MAC to guarantee message integrity and authentication.

4.3.4.4 Key Management

All kinds of security services are based on different encryption algorithms, and the security of the algorithm relies on the security of the key, so key management involves the security and reliability of all security services. Key management involves many aspects including key generation, distribution, utilization, updating, storage, backup, and destruction.[21]

There are two kinds of keys in aircraft and ground communication applications: one is the asymmetric key used for digital signature or certification and the other is the symmetric key used for link encryption and integrity services.

4.3.4.4.1 PKI-Based Key Management

Elliptic curve cryptographic (ECC) algorithms are employed to perform secure session initiation and key establishment between communicating peer AMS aircraft and ground entities in the case that AMS is implemented using public/private keys. Public key cryptography uses a pair of asymmetric keys, which are related mathematically, and one-way functions, and the important characteristic of the keys is that it is easy to compute using one of the keys but very difficult to solve without knowing the related key. One of the keys must be kept private and protected from compromise to ensure security, while another one may be distributed publicly. Although one key is public, it is very difficult to derive the private key by knowing just the public key that is made by the mathematical properties of cryptographic algorithms such as ECC.[12]

PKI provides life-cycle management of the public/private keys to ensure secure sessions for AMS entities. CA, a trusted PKI entity, issues public key certificates to AMS entities. During the issuance, CA digitally signs the content of the public key certificate and the public key certificate binds the public portion of the key pair with the identity of the AMS entity that uses the key pair.

Since there is no sensitive cryptographic information in the public key certificate and the public key is protected by the CA digital signature from modification, the CA may publish the public key certificate via an unprotected channel. Before using public key certificates, AMS entities must verify the CA's digital signature to ensure that the certificate has been certified and has not been modified during the storage or distribution.

Since the public key certificates in asymmetric cryptography are permitted to be distributed freely, PKI-based key management is particularly suited for the following situations:

- It is practical and feasible to preplace and/or maintain the pair-wise shared-secret keys.
- The quantity of entities increases over time. So scalability is important.
- Security services must interoperate across multiple, heterogeneous organizations. For example, AMS DSP-based security is that one or more service providers offer AMS security services to multiple airline organizations.
- There is a need to know the identity of the entity with which secure communication is being established, for example, an aircraft crew needs to authenticate the identity of a ground entity and vice versa.

4.3.4.4.2 Shared Secret Key Management

Symmetric algorithms are employed to perform secure session initiation and key establishment between communicating peer AMS aircraft and ground entities when AMS is implemented using a shared secret key. Each AMS entity is required to share a secret key in the symmetric algorithms. Also, each AMS entity must protect the shared secret key from compromise to ensure security. If the key is compromised for one entity, then it is equally compromised for all entities that share the same key.[12]

A key management infrastructure (KMI) provides the necessary life-cycle management of the shared secret keys to ensure secure session for AMS entities. The KMI is responsible for generation and distribution of a shared secret key to each pair of AMS entities. The shared secret key is distributed to each AMS entity through a secure communication channel independent of the ACARS datalink.

Shared secret–based key management is particularly suited for the situations as follows:

- The quantity of entities is small and generally does not change over time. So scalability is not important.

- Security services must be within an organization for interoperability among entities. For example, AMS end-to-end security is that one airline manages all of the aircraft entities and ground entities that must communicate securely.

- It is practical and feasible to preplace and/or maintain the pair-wise shared-secret keys.

- There is no need to know the identity of the entity with which secure communication is being established.

4.3.4.5 Payload Encoding

ACARS is a character-oriented communication system, but almost all encryption algorithms are bit-oriented until now. Hence, the ACARS message should be converted to bit stream before it is encrypted by encryption algorithm. However, the characters used in the ACARS message are all noncontrol characters in the ISO 5 character set, and they rarely use lowercase letters. Meanwhile, the information coding before transmission will increase the length of the message. As a result, it is really necessary to load coding for the ACARS message.

The ACARS message should first be converted from character to bit using an encryption algorithm. ACARS uses the ASCII character set apart from the control character that does not appear in the user data stream; we have organized the most frequently used 64 characters as the coding in Table 4.7 after analyzing a large number of ACARS messages. Based on the table, an 8-bit character can be converted to 6-bits, and then they are connected as bit stream to decrease the effective load by 25%.

As to the characters apart from the coding table, we delete the most significant bit 0 to reduce the 8-bit character to a 7-bit character stream, and the effective load can be decreased by 12.5%. The length of the bit stream will be an integral multiple of eight no matter which coding method is selected. Otherwise, it needs to pad with padding bit by bit.

In the recipient, the original ACARS message data are available after the bit stream being decoded is anticoded based on the coding table.

This coding way not only converts the ACARS message to bit stream for data encryption, but also allows for information coding later to be more effective and practical.

4.3.4.6 Information Coding

All ACARS security services, including data encryption and the ACARS message with MAC, are required to undergo information coding before they are transferred by the ACARS network. Information coding also uses a 64-bit character coding table, and the information coding table is basically the same as the load coding table, apart from the character LF ($0 \times 0A$) and CR ($0 \times 0D$) that are replaced by "[" ($0 \times 5B$) and "]" ($0 \times 5D$).

TABLE 4.7

Coding Table of 6 Bits Effective Load

Character		6 Bits	Character		6 Bits	Character		6 Bits	Character		6 Bits
ASCII	HEX	Coding	ASCII	HEX	Coding	ASCII	HEX	Coding	ASCII	HEX	Coding
SP	0×20	000000	0	0×30	010000	@	0×40	100000	P	0×50	110000
!	0×21	000001	1	0×31	010001	A	0×41	100001	Q	0×51	110001
"	0×22	000010	2	0×32	010010	B	0×42	100010	R	0×52	110010
#	0×23	000011	3	0×33	010011	C	0×43	100011	S	0×53	110011
$	0×24	000100	4	0×34	010100	D	0×44	100100	T	0×54	110100
%	0×25	000101	5	0×35	010101	E	0×45	100101	U	0×55	110101
&	0×26	000110	6	0×36	010110	F	0×46	100110	V	0×56	110110
'	0×27	000111	7	0×37	010111	G	0×47	100111	W	0×57	110111
(	0×28	001000	8	0×38	011000	H	0×48	101000	X	0×58	111000
)	0×29	001001	9	0×39	011001	I	0×49	101001	Y	0×59	111001
*	0×2A	001010	:	0×3A	011010	J	0×4A	101010	Z	0×5A	111010
+	0×2B	001011	;	0×3B	011011	K	0×4B	101011	LF	0×0A	111011
,	0×2C	001100	<	0×3C	011100	L	0×4C	101100	\	0×5C	111100
-	0×2D	001101	=	0×3D	011101	M	0×4D	101101	CR	0×0D	111101
.	0×2E	001110	>	0×3E	011110	N	0×4E	101110	^	0×5E	111110
/	0×2F	001111	?	0×3F	011111	O	0×4F	101111	|	0×7C	111111

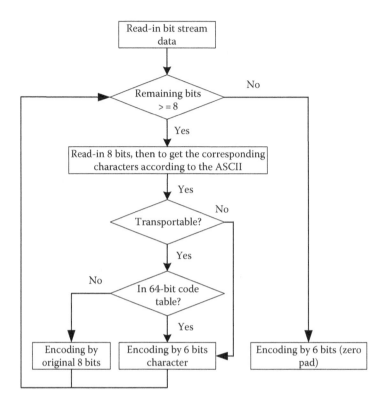

FIGURE 4.7
Information coding process.

First, we convert the message to bit stream. Eight bits are read from high bit, and it is judged whether they are characters that could be transferred. If the answer is *no,* then we code the first 6 bits to an 8-bit character based on the information table. If the answer is *yes,* we continue to judge whether the messages are in the information table. If the answer is *yes,* we code the first 6 bits to an 8-bit character based on the information coding table. If the answer is *no,* we code the original 8 bit. According to such steps over and over again, until the remaining number of bits are equal to or less than 6 bits, we code the message based on the information in the coding table (zero is padded while the length of the message is shorter than 6 bits). Finally, we get the result. The process is shown in Figure 4.7.

After information coding, the final message consists of all transportable ACARS characters and the result could be transferred by the ACARS network.

4.3.4.7 Data Compression

The purpose of data coding and data compression is to reduce the effect of the security load as much as possible. The AMS standard includes coding and compression for the effective load, which lowers the length of the AMS information. Besides, the data compression program and the AMS security mechanism produce arbitrary binary bit stream and an information coding mechanism offers a harmless method to convert the bit-oriented information stream to a character-oriented information stream so as to transfer to the ACARS network.

AMS offers the mechanism of a compression algorithm. Both sides of the communication offer a compression algorithm supported by themselves, and an optimal algorithm is finally chosen by negotiation. The compression rate is determined by the original information. Generally speaking, compared to shorter and more random information, the longer information with a higher repetition rate has a higher information compression rate. AMS offers two default compression algorithms, DMC and DEFLATE.

4.3.5 Conclusion

This chapter analyzes some main security threats for the ACARS datalink and offers detailed solutions to a variety of threats. Meanwhile, it introduces varieties of algorithms used in the solution, such as encryption, authentication, key exchange, and coding and decoding.

References

1. Z. Xuebing and M. Yuwen. VHF air/ground datalink system and ARINC 618 protocol. *Aviation Maintenance and Engineering*, 2002, 2(1):41–43.
2. ARINC. Datalink security part 1–ACARS message security, ARINC specification 823P1. Annapolis, MD: Aeronautical Radio, Inc., December 10, 2007.
3. X. Xiaogang. *The Research and Application of Data-Link Technology in Tower Control*. Beijing, China: Beihang University, 2004.
4. H. Junxiang. Research on ACARS data process. *Journal of Civil Aviation University of China*, 2007, 25(1):1–3.
5. X. Wenhui. Research on the realization technology of ACARS. *Avionics Technology*, 1995, 26(2):25–28.
6. ARINC. ARINC specification 618-5: Air/ground character-oriented protocol specification. Annapolis, MD: Aeronautical Radio, Inc., 2000.
7. G. Hong, J. Jun, and X. Wenyan. Parameter identifying of the ACARS message. *Avionics Technology*, 2006, 37(4):6–11.
8. ARINC. ARINC specification 620-4: Data link ground system standard and interface—Specification (DGSS/IS). Annapolis, MD: Aeronautical Radio, Inc., 1999.
9. H. Jia. Air/ground data link system and ARINC 620 protocol. *Aviation Maintenance and Engineering*, 2004, 5(1): 33–35.
10. W. Xiaolin, Z. Xuejun, and H. Jia. Security communications in ACARS data link. *Avionics Technology*, 2003, 34(zl):95–100.
11. C. Hao, H. Yisheng, J. Tao, X. Yongchun, and W. Yunbing. Status and development of the data link security techniques. *Electronic Warfare*, 2006, 111(6):37–41.
12. ARINC. Datalink security part 1—Key management, ARINC specification 823P2. Annapolis, MA, ARINC, March 10, 2008.
13. S. Tao, M. Hongguang, and X. Wentong. Research and application on network data encryption. *Computer Engineering and Application*, 2002, 38(19):156–158.
14. National Institute of Standards and Technology. Advanced encryption standard, draft federal information processing standard. Gaithersburg, MD: National Institute of Standards and Technology, 2001.
15. A. Kahate. *Cryptography and Network Security*. Beijing, China: Tsinghua University Press.
16. H. Krawcyzk, M. Bellare, and R. Canetti. HMAC: Keyed hashing for message authentication. Internet Engineering Task Force, Internet RFC 2104, 1997.

17. W. Hongxia, L. Saiqun. A mechanism for message authentication using HMAC-SHA1. *Journal of Shanxi Teacher's University (Natural Science Edition)*, 2005, 19(1):30–33.
18. L. Zhigui, Y. Lichun, P. Jie et al. The system of digital signature authentication based on PKI. *Application and Research of Computers*, 2004, 9:58–160.
19. V. Patel and T. McParland, Public key infrastructure for air traffic management systems. *20th Digital Aviation Systems Conference*, Daytona Beach, FL, October 2001.
20. American National Standards Institute. Public key cryptography for the financial services industry—The elliptic curve digital signature algorithm (ECDSA), ANSI X9.62. American National Standards Institute, 1998.
21. Y. Zhijun. Distribution and management of online security and key. *Journal of Taiyuan Teachers College (Natural Science Edition)*, 2004, 3(3): 22–24.

5

VDL2 Key Technology and Simulation

Gao Lin

CONTENTS

5.1 VDL General Introduction

The very high-frequency digital link (VDL) is an air–ground subnetwork that may be applied in supporting data communication across aeronautical telecommunication network (ATN) aircraft-based application processes and the peer ground-based processes. Presently, the VDL has been recommended as the major method of air traffic management (ATM) in the transition to the future ATN and has been deployed in many areas such as central Europe and North America.

5.1.1 Transition from ACARS to the VDL

Before the advent of the data link system, all communications between the ground staff and flight crew were only through voice by means of very high-frequency or high-frequency voice radio communication. To reduce the workload and improve data integrity, Aircraft Communication Addressing and Reporting System (ACARS) was proposed as the first aeronautical air–ground data link in the 1970s. ACARS was first researched and developed by ARINC. Afterward, ARINC and SITA have become the two main providers of the ACARS system in the world. For a long time, it has been a very popular aeronautical data link and has become the dominant air–ground communication mode worldwide.[1]

ACARS works in very high frequency (VHF) with 25 kHz channel band. The modulation scheme of the system is amplitude modulation minimum shift keying (AM-MSK). It has no forward error correction (FEC). The channel access scheme is nonpersistent carrier sense multiple access (CSMA). It is character oriented and can communicate only at 2.4 kbps.

With the development of air service, ACARS has become unable to satisfy the high-capacity need of aviation communication. VDL technology emerges as the time requires. It can arrive at far more excellent performance than ACARS. For example, the VDL2, as a typical VDL mode, has a 13 times higher transmission rate than ACARS.

The VDL2 is a bit-oriented packet communication system with 25 kHz channel separation. The modulation scheme of the system is differential 8 phase-shift keying (D8PSK). The VDL2 has a FEC function with Reed–Solomon (RS) code. The channel access scheme is p-persistent CSMA. It can get a 3.15 kbps transmission bit rate.

A performance comparison between the VDL2 and ACARS has been made by simulation.[2] In the simulation, the assumptions were made that the message data length and data generation interval, respectively, followed uniform distribution and Pareto distribution according to flight data analysis. Also, the load condition approximated the actual ACARS data ratio (Table 5.1). The conclusions are as follows:

1. When the data traffic load generated from an aircraft equaled the load directed to the aircraft, the VDL2 was able to process 4.6 times more congested load than ACARS.

2. When the data traffic load generated from an aircraft was five times higher than the load directed to the aircraft, the VDL2 was able to process 8.8 times more congested load than ACARS.

Besides the VDL2, modes 3 and 4 also have their advantages over ACARC in performance. Compared to the VDL, there are three defaults in ACARS, which restrict the system performance. First, AM mode determines that the frequency utilization rate is relatively low, and the system is susceptible to radio-frequency (RF) interference. Second, the communication technology of the ACARS data link, as the character-oriented system, is less flexible than bit-oriented one. Third, not like the VDL, ACARS is not compatible with open systems interconnection (OSI), which is adopted by the ATN.

Under the growth in air traffic and the development of the ATM services, the VDL is attracting more attention of civil aviation organizations and institutions in major countries and regions of the world. The VDL concerned test and deployment works are being carried out systematically. The aeronautical data link is nonexperiencing a transition from ACARS to the next-generation data link—the VDL.

5.1.2 VDL Modes

As yet, the ICAO has approved four VHF modes as the ATN-supporting data link and has written them in Annex 10.[3] Among them, the VDL1 is the intermediate path from the existing ACARS to the VHF data link modes and has less practical value. Accordingly, only three modes are described and discussed in the subsequent sections.

5.1.2.1 VDL2 Character

The VDL2 is the air–ground data link designated by ICAO aviation mobile communications experts group (Aeronautical Mobile Communications panel [AMCP]) in 1997. There are many countries involved in its development and construction. Europe is the main VDL2 supporter and implementer. They began to test the feasibility of using the VDL2 data link in 2001 and have carried out the Link2000+ project relying on Eurocontrol. Japan is another VDL2 supporter and has built two ground stations (GSs) by early 2004.[4]

In the VDL2, the channel spacing is 25 kHz, and the modulation scheme is D8PSK. In the data link layer, the VDL2 applies CSMA in the medium access control (MAC) sublayer. In the data link service (DLS) sublayer, it specifies the aviation VHF link control (AVLC) protocol, which is derived and developed from high-level data link control (HDLC) protocol. The VDL2 transmission rate is 31.5 kbps. (For details of the protocol and key technology of the VDL2, refer to the next two sections in the chapter.)

The VDL2 can provide air–ground communication and uplink broadcast service. It can provide Aeronautical Operational Communication, as well as be used for CPDLC, ATIS, ATC, etc., applications. However, the VDL2 does not support voice so far. Also, it does not provide priority function, which increases the information delay in the high offered load circumstance and makes it unsuitable for emergency use.

TABLE 5.1

Comparison between ACARS and the VDL2

System	Modulation	Channel Space (kHz)	Information Unit	Information Type	Error Correction	Access Scheme	Supported Protocol	Transmission Bit Rate (kbps)
ACARS	AM-MSK	25	Character	Date	No	Nonpersistent CSMA	ACARS	2.4
VDL2	D8PSK	25	Bit	Date	Yes	p-persistent CSMA	ATN ACARS	31.5

5.1.2.2 VDL3 Character

The VDL3 is the data link that FAA energetically pushes for implementation in the United States. FAA intends to use multimode digital radio to gradually replace the existing ground VHF/UHF stations in the United States.

The VDL3 can transmit both data and voice. As with mode 2, the VDL3 applies 25 kHz channel spacing and specifies a D8PSK modulation scheme operating at 31.5 kbps. The voice coding transmission rate is 4.8 kbps (normal voice) or 4 kbps (truncated voice).

In the VDL3, the MAC sublayer applies time division multiple access (TDMA) in which slot usage for data transmission is controlled by the GS. Several configurations are applied with the standard four-slot per frame and long-range three-slot per frame timing structures to accommodate different amounts of voice and data traffic. In the standard range configurations, four 30 ms time slots, denoted as A, B, C, and D, make up a MAC frame. A MAC cycle consists of two MAC frames, even and odd, with a duration of 240 ms. Time slots are subdivided into Logical Burst Access Channels (LBACs) to provide management capabilities in the voice and data slots.

The VDL3 supports four levels of priority. Based on priority, the ground decides when to grant access to the link. Reservation responses are sent in the uplink management (M) burst in slot A of the odd frame. If the ground is unable to immediately grant the request when it is received, it responds with a request acknowledgment (RACK). Otherwise, it will indicate which slot in the next MAC cycle may be used to begin the transfer, as well as whether slot D may be used or if it is reserved for voice. Once access has been granted to the link, a station will transmit starting in the indicated slot and apply consecutive data slots until the transmission is completed. Transmissions are limited to a maximum of 15 consecutive slots. In the event that a voice reservation is also granted, only slots B and C will be used for data transmission.

DLE uses an acknowledged connectionless protocol (ACLP) that adds only 6 bytes of header information to each network frame. Upon sending a frame, DLE waits for an acknowledgment before sending the next frame. The MAC sublayer notifies the DLE when an ACK is not received at the expected time. Frame grouping is allowed at the DLS sublayer, in which frames of the same priority and destination may be sent together. Optionally, a frame of lower priority may also be included in the group as long as it does not require additional slots to transmit. An ACK will acknowledge all frames in the group. If the T1 timer expires before an acknowledgment is received and retransmission occurs, and a higher priority packet is queued, the retransmission of the unacknowledged group will be delayed.

Downlink acknowledgments created by the DLE are automatically converted into an ACK burst and scheduled for transmission in the M burst LBAC after the last data segment of one MAC cycle is sent. The GS recreates the DLEACK from the ACK burst and passes it to the DLE. Uplink acknowledgments are sent in normal data bursts, but with an expedited priority that takes precedence over any ungranted data transmission.

The VDL3 can support voice transmission and can provide air–ground communication and uplink broadcast, etc., ATN services. At the same time, it is fully compatible with the ATN in data transmission.

5.1.2.3 VDL4 Character

The VDL4 is the scheme proposed by Sweden. With the global navigation satellite system (GNSS) information timing and the bit-oriented protocol, the VDL4 can support multiple communication services for the ATN and even can provide more than the ATN services, such as ADS-B. At present, Eurocontrol is committed to the development of the VDL4 for the communication and surveillance services, including the system standard, frequency, and structure.

As with modes 2 and 3, the VDL4 channel spacing is 25 kHz, but the physical layer uses a Gaussian-filtered frequency shift keying (GFSK) modulation operating at 19.2 kbps. MAC uses a self-organizing time division multiple access (STDMA) mechanism. Time is segmented into 13.3 ms timeslots, with 75 slots per second. A superframe lasts for 60 s and contains 4500 slots. Aircraft timing is primarily based on GNSS, but alternatively, timing can be derived from GSs or other sources, including other aircraft. This allows the system to continue operating even when communication with the primary source fails.

The MAC burst for mode 4 requires more information to be contained within it than the other modes. Mode 4 bursts contain a start and an end flag, reservation information, source address, message, and a cyclic redundancy check (CRC). The size of the burst header depends on the type of message being transmitted and the type of reservation being placed.

The VDL4 receivers are intended to be multichannel. Two global signaling channels are defined for mode 4. Aircraft are expected to transmit position information, such as in ADS-B messages, alternately on these two channels. In areas where traffic levels are high, the aircraft can be directed to local channels for use.

The VDL4 has two operational modes: autonomous reporting and directed reporting. When operating under autonomous reporting, each aircraft chooses its own slots for transmission, whereas in directed reporting, the GS determines which slots aircraft may use.

Slot selection is handled by the VSS. When a reservation is to be placed, the VSS identifies the first Q4 available slots in the range specified by the application that meet the length requirement. In the event that less than Q4 available slots are found in the specified range, a mechanism exists to select previously reserved frames from distant aircraft communicating with distant stations. This *Robin Hood* protocol allows for slot reuse if certain conditions are met. From the available list of slots, the VSS randomly selects one for use.

Several reservation mechanisms exist for the VDL4, such as periodic broadcast, incremental broadcast, unicast request, information transfer request, directed request, superframe block, and second block reservation protocols. The unicast reservation allows for a one-way transfer. A station can reserve a block of slots on behalf of another station. The information transfer protocol, however, is a two-way communication. A station reserves slots for another station to use, and it also reserves a single slot for itself to send an acknowledgment.

The unicast reservation utilizes a VSS-based retransmission mechanism. If a response is not received from the peer in the slot reserved for the transfer, the retransmission procedures will be invoked. Retransmitted frames access the link by random access. The retransmission delay is dynamically calculated based on slot utilization and the number of retransmission attempts. The value of Q5 is randomly set, according to the standards, between a minimum and a maximum value, where the maximum is determined by the number of retransmissions and the minimum by default is 0.

Information transfers do not use the VSS for retransmission; the retransmission is the responsibility of the DLS sublayer. Upon indication that an information transfer is not successful, a DLE will restart the transmission process. The DLE uses a negotiated setup connection-oriented protocol (NSCOP) for air-to-ground communications and a zero overhead connection-oriented protocol (ZOCOP) for air-to-air communications. The DLS supports 16 levels of priority, enough to map each of the 15 levels of priority specified by the ATN. The larger burst format at the MAC sublayer allows for a small DLE frame header to be used. For data packets, the header is only 2 bytes in length.

The DLS will use either the short or long transmission procedures depending on the length of the frame. In the short procedure, the DLS sends a data frame to the VSS with instructions to include a unicast reservation for a single slot with the transmission. The VSS will place a reservation for the data frame in an existing transmission, if possible; otherwise, it will specify that the frame be transmitted by random access. When the frame is received, an acknowledgment is created by the peer DLS and is transmitted back to the source. The acknowledgment will be placed in the slot reserved by the unicast reservation.

For a long transmission, the DLS issues a request-to-send (RTS) frame that includes the length of the data transmission. The RTS is sent similarly to the data frame of the short transmission, with a unicast reservation for a single slot for the peer. The peer, upon receiving an RTS, responds with a clear-to-send (CTS). The CTS is sent in the frame reserved for it and includes an information transfer reservation for the data length specified in the RTS. The information transfer reservation reserves slots for the data frame as well as a slot for the acknowledgment. The DLS will send the frame when it receives the CTS, and the peer will respond with an acknowledgment.

DLE allows linking of transmissions such that data and RTS frames can be combined in a single transmission. In these cases, the ACK and CTS responses will also be combined. When transmissions are linked in this way, all transmissions after the initial RTS will be by reserved access.

The VDL4 is fully compatible with the ATN network. It can realize the air–ground communication, uplink and downlink broadcast, air-to-air broadcast, etc., ATN services. It also can provide more than the ATN services, such as ADS-B and air-to-air communication. As its limitation, the VDL4 does not support voice.

5.1.3 VDL Modes Comparison

We can compare the modes from the aspects of system performance, support service, technology compatibility, etc.

1. The VDL2 has the highest capacity, which can be valued by the supported aircraft (AC) number under the delay required by corresponding regulation. The VDL MASPS states that the 95th percentile subnetwork delay should be no more than 3.5 s for mode 2. For mode 3, the maximum 95th percentile subnetwork delay for high-priority traffic of 192 bits or less is 1 s, and the maximum 99.9th percentile subnetwork delay is 5 s, which is shorter and stricter than the VDL2 due to the priority support functionality.[5] Since mode 4 has only been standardized for use in the ATN for surveillance and not as a communications data link, a maximum limit does not exist and is substituted by the mode 3 definition. Under the delay limits, simulations determined that Mode 2 is capable of bearing 3.3 kbps load, which is equivalent to the load from approximately 130 aircraft. Compared to mode 2, modes 3 and 4 have far less load with only mode 3 supporting 0.5 kbps and mode 4 supporting 0.8 kbps, which are equivalent to the load from approximately 20 and 30 aircraft, respectively.[6]

TABLE 5.2

VDL Modes' Comparison

Mode	Modulation	Media Access	Transfer Control	Priority	Voice Support	Estimated Max Load (kbps)	Transmission Bit Rate (kbps)
VDL2	D8PSK	CSMA	AVLC	No	No	3.3	31.5
VDL3	D8PSK	TDMA	ACLP	4 levels	Yes	0.5	31.5
VDL4	GFSK	STDMA	NSCOP/ZOCOP	15 levels	No	0.8	19.2

2. In the terminal and en route domain simulations, the VDL2 delay is on the whole better than the VDL3 and almost equivalent to the VDL4. The experiments were under the same traffic model with typical services and message rates. With the default parameters regulated by the corresponding protocols, the three modes result in similar average uplink delays. In the downlink, modes 2 and 4 can provide lower average 95th percentile delays than mode 3. With the optimized parameters, mode 2 has 99.9th percentile delays rivaling those of modes 3 and 4 in the en route domain and is better than modes 3 and 4 in the terminal domain.

3. In service aspect, modes 2 and 4 cannot support voice, while mode 3 can support both data and voice service. Mode 4 can supply diverse date service for both ATN and non-ATN, for example, the VDL4 can support ADS-B and air-to-air communication, which are the services more than the ATN. On the other hand, modes 3 and 4 support priority and can be used for safety-of-flight messages. Because of CSMA implementation, the VDL2 cannot provide prioritized link access and is intended to communicate noncritical information.

4. Different from modes 3 and 4, the VDL2 not only support the ATN, but also ACARS. The VDL2 can support messages in ACARS format and can deal with ACARS protocol by ACARS over AVLC technology. The VDL2 also has more similarity in implementation with ACARS than other modes, for example, the two systems implement CSMA for media access. Due to these reasons mentioned, the VDL2 has the best compatibility with ACARS, which has the largest data link share in the current aviation industry.

From the VDL mode comparisons, we can know that the VDL2 has the best performance as a whole (Table 5.2). It can achieve the highest capacity and almost has the shortest delay. Besides, the VDL2 system has the best compatibility with ACARS among the VDLS, which can benefit and smooth the transition from the current system to the future ATN. At the same time, the deficiency of the VDL2 service (e.g., it does not support voice and priority) can be made up by using another communication system, such as aeronautical mobile satellite system. In summary, the VDL2 is the most hopeful VHF data link to succeed ACARS.

5.2 VDL Protocol

As an ATN-compatible data link, the VDL2 has been approved by the ICAO. The VDL2-related protocol is stated in Standards and Recommended Practices (SARPs), which were developed by the AMCP[7] and introduced in the ICAO Annex 10.[8]

5.2.1 Stacks Structure

The international aviation community is expected to adhere to the separation of communication functions as specified in the OSI reference model developed by the International Organization for Standards (ISO). The OSI reference model permits the development of open communications protocols as a layered architecture comprising seven functional separate layers.

As an entire network, the ATN conforms to the OSI reference model and has all of its seven layers, including low-level three layers and high-level four layers. The VDL2, which is the ATN communication subnet and is fully compatible with the ATN, also conforms to the OSI reference model but only has the low level. It constitutes the first step toward a fully OSI-compatible protocol stack (Figure 5.1).

The ATN can be divided into ground subnetwork, avionics subnetwork, and air–ground subnetwork. Among them, ground subnetwork and avionics subnetwork belong to fixed data links, which, respectively, realize in aircraft (AC) and ground; air–ground subnetwork belongs to mobile data link. The subnetworks can be regarded as being made up of end system (ES) and intermediate system (IS), which are two elemental systems abstracted from real equipment. ES can be defined as the computer system that executes end-user application programs and communicates to IS or possibly to other ESs. ES can exist in both avionics subnetwork (e.g., display control system of aircraft cockpit) and ground subnetwork (e.g., the ATM system of GS). IS can be defined as the computer system that relays and routes network protocol data unit (NPDU) to other ISs or ESs. Network routers are the typical IS equipments. ES may implement the total ATN seven-layer stack as required, while IS implements only the low-level layers.

The VDL subnetwork realizes the communication between avionics subnetwork and ground subnetwork. It locates between the ATN ISs (avionics subnetwork router and ground subnetwork router) and implements only the lower three layers of OSI reference model.

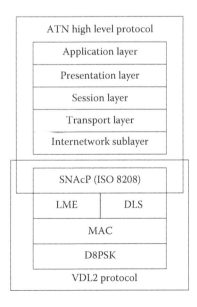

FIGURE 5.1
The VDL2 protocol stack in the ATN.

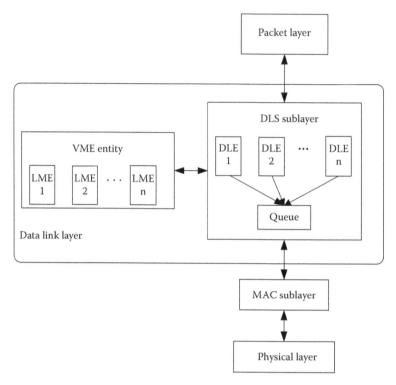

FIGURE 5.2
The VDL2 system architecture.

For the VDL2, the physical layer (at the bottom of system architecture) specifies a D8PSK modulation scheme and operates at a bit rate of 31.5 kbps. All transmitted physical frames contain a training sequence (Figure 5.2).

Above the physical layer is data link layer, which can be divided into MAC sublayer, DLS sublayer, and the VDL management entity (VME). MAC is the interface between physical layer and data link layer. It is governed by a p-persistent CSMA protocol, in which access attempts are made only when the channel is sensed idle. During an access attempt, the station will transmit with probability p or back off for $TM1$ seconds with a probability $(1 - p)$. The maximum number of access attempts is bounded, and after a maximum number of failed access attempts, the MAC will transmit the packet as soon as the channel becomes idle. This algorithm attempts to reduce the number of collisions while minimizing the medium access delay.

The DLS employs a connection-oriented protocol (COP) known as aviation VHF link control (AVLC). The AVLC protocol is a sliding window protocol with multiselective reject functionality. It uses a dynamically calculated retransmission delay T1 based on the channel utilization and the number of retransmissions. The value of T1 is randomly set between a minimum and a maximum value, where the maximum is determined by the number of retransmissions. The DLS can support more than one data link entity (DLE). All frames sent from DLEs are placed in an active DLS queue. The queue processes frames for acknowledgment and redundancy. The queue guarantees that redundant or duplicate packets are not contained in the queue. When a frame is queued, its acknowledgment is checked, and all other frames to the same destination are updated to reflect that acknowledgment. Incoming frames are also checked, and any queued frames that the incoming frame acknowledges are removed. When the access to the medium is successful, link control frames are transmitted in preference to user data (UD). When the channel is accessed, more than one DLE frame may be included in a single physical layer transmission.

VME provides link management functions, including link setup/release and handover. It can support more than one link management entity (LME). Packet layer (subnet layer) is mainly responsible for packet segment/reassembly, packet switching and error recovery, subnet connection management, etc., functions.

5.2.2 Physical Layer

In baseband processing, the VDL2 applies the Nyquist first criterion with the raised cosine roll-off filter ($\alpha = 0.6$) to eliminate intersymbol interference. In band processing, the VDL2 applies D8PSK modulation scheme with the message rate of 31.5 kbps. In error correction, the VDL2 adopts the FEC mechanism and applies RS channel coding technology. According to SARPs, scrambling and interleaving are also recommended to implement in physical layer. The former can improve the synchronization performance of data stream; the latter can enhance the capacity of anti-burst error by means of transforming channel.

In frequency allocation, the VDL2 assignable frequency covers the band 118–136.975 MHz. The separation between assignable frequencies (channel spacing) is 25 kHz, which can accommodate up to 760 channels. Frequency 136.975 MHz shall be reserved as a worldwide common signaling channel (CSC) for the VDL Mode 2.

5.2.2.1 Architecture

As the stack bottom, the physical layer provides a physical link, which can realize transparent transmission for communication. In the communication process between the peers, the data link layer at the transmitter side, from high to low, transfers control signaling and UD to the physical layer through service primitives. The physical layer will implement coding/modulating onto the received message and allocate physical channel according to upper layer request, then transmit the processed message to the physical layer in the peer entity. At the receiver side, the physical layer transfers the received data to data link layer through service primitives. In short, the physical layer provides service for the data link layer in the local entity, and the peer layers realize physical layer protocol by the primitive-based services.

5.2.2.2 Functionality

The VDL2 physical layer can provide transceiver frequency control, bit exchange over radio, notification, etc., functions, as follows:

1. *Transceiver frequency control*: According to the link layer requests, selects transmitting and receiving frequency.
2. *Notification*: The GS notifies signal quality by signal quality parameter (SQP), which is the basic for handover decision and link selection.

3. *Data transmitting*: Encodes and modulates the data from the data link layer, transmits the data via RF channel.

4. *Data receiving*: Via RF channel, receives the data from the peer transmitter, then decodes and demodulates the data. It is the inverse function of data transmitting.

5. *Bit synchronization*: The receiver extracts the synchronization information from the transmitter and ensures the right time in data decision.

5.2.2.3 Physical Frame Structure

The VDL2 physical frame can be divided into six domains, the first five segments of which construct training sequence (Table 5.3).

Each segment of training sequence has the meaning as follows:

1. *Transmitter ramp-up and power stabilization*: The purpose of the segment, called the ramp-up, is to provide for transmitter power stabilization and receiver AGC settling, and it shall immediately precede the first symbol of the unique word. The duration of the ramp-up shall be five symbol periods. The time reference point (t) is the center of the first unique word symbol that occurs half a symbol period after the end of the ramp-up. Conversely stated, the beginning of the ramp-up starts at $t = -5.5$ symbol periods. The transmitted power shall be less than -40 dBc prior to time $t = -5.5$ symbol periods. The ramp-up shall provide that at time $t = -3.0$ symbol periods, the transmitted power is 90% of the manufacturer's stated output power or greater. Regardless of the method used to implement (or truncate) the raised cosine filter, the output of the transmitter between times $t = -3.0$ and $t = -0.5$ will appear as if "000" symbols.

2. *Synchronization code*: The segment consists of the unique 48-bit word: 000 010 011 110 000 001 101 110 001 100 011 111 101 111 100 010, which implements the symbol synchronization function.

3. *Reserved symbol*: The segment consists of the single symbol representing 000 and is reserved for future definition.

4. *Transmission length*: To allow the receiver to determine the length of the final RS block, the transmitter shall send a 17-bit word, from least significant bit (lsb) to most significant bit (msb), indicating the total number of data bits that follow the header FEC. From the field, we can know that the maximum length of physical frame is $2^{17} - 1 = 131,071$ bits.

5. *Header FEC*: To correct bit errors in the header, a (25, 20) block code shall be computed over the reserved symbol and the transmission length segments. The

TABLE 5.3

VDL2 Physical Frame Structure

Transmitter Ramp-Up and Power Stabilization	Synchronization Code	Reserved Symbol	Transmission Length	Header FEC	Message
5 symbols	48 bits	3 bits	17 bits	5 bits	Variable

encoder shall accept the header in the bit sequence that is being transmitted. The parity bits to be transmitted shall be generated by the following equation, where P is the parity symbol, R is the reserved symbol, TL is the transmission length symbol, H is the parity matrix, and T represents the matrix transpose function.

$$\begin{bmatrix} P_1 & P_2 & P_3 & P_4 & P_5 \end{bmatrix} = \begin{bmatrix} R_1 & R_2 & R_3 & TL_1 & \cdots & TL_{17} \end{bmatrix} P^T \tag{5.1}$$

$$P = \begin{bmatrix} 0\,0\,0\,0\,0\,0\,0\,0\,1\,1\,1\,1\,1\,1\,1\,1\,1\,1\,1\,1 \\ 0\,0\,1\,1\,1\,1\,1\,1\,0\,0\,0\,0\,1\,1\,1\,1\,1\,1\,1\,1 \\ 1\,1\,0\,0\,0\,1\,1\,1\,0\,0\,1\,1\,0\,0\,0\,0\,1\,1\,1\,1 \\ 1\,1\,0\,1\,1\,0\,1\,1\,0\,1\,0\,1\,0\,0\,1\,1\,0\,0\,1\,1 \\ 0\,1\,1\,0\,1\,0\,0\,1\,1\,1\,1\,0\,0\,1\,0\,1\,0\,1\,0\,1 \end{bmatrix}$$

5.2.3 Data Link Layer

The VDL2 data link layer contains three parts: MAC sublayer, DLS sublayer, and the VME.

5.2.3.1 Media Access Control

MAC sublayer applies P-CSMA and implements physical channel access and control.

5.2.3.1.1 Architecture

MAC sublayer shall accept the services from its lower layer, for example, it can acquire channel idle/busy state from physical layer and use the bit in P-CSMA algorithm. It shall provide the transparent acquisition for shared communication path and supply services to the upper layer. As a whole, it can be regarded as the interface between the physical layer and the data link layer.

5.2.3.1.2 Functionality

1. *Data transmitting/receiving*: MAC sublayer provides transparent transmission functionality for upstream data and supplies P-CSMA-based access control for downstream data.
   a. *Transmitting process*: MAC transfers AVLC frame from the DLS to physical layer, which is a downward communication.
   b. *Receiving process*: MAC transfers AVLC frame from physical layer to the DLS, which is an upward communication.
2. *Multiple access control*: Based on P-CSMA algorithm, MAC sublayer can administrate and allocate the VHF physical media and can realize channel sharing and multiplexing.
3. *Channel congestion notification*: When the channel is congested, MAC sublayer shall send the channel congestion notification to VME, which can notify VME to initiate a handoff.

5.2.3.2 Data Link Service

The DLS sublayer implements the AVLC protocol, mainly realizing frame sequencing, error detection and recovery, and GS identification functions.

5.2.3.2.1 Architecture

The DLS sublayer is responsible for the organization and the transmission of packets on the data link, implementing AVLC, which is derived from the HDLC.

The DLS sublayer shall build a DLE link for any point-to-point communication. A supervision relationship shall come into being between a local DLE and its associated remote DLE, implementing the peer-to-peer communication control. At the same time, DLE is also responsible for notifying packet-error information to VME.

5.2.3.2.2 Functionality

The main functions of the DLS sublayer are as follows:

1. *Frame sequence reception*: In accordance with the serial number provisions, the DLS sublayer can decide and receive AVLC frame.
2. *Flow control*: Receiving side shall control the data transmission flow by sending RR frame (RNR frame is not currently used) to the peer.
3. *Error detection*: The DLS sublayer has the ability to detect and abandon the error frame caused by transmission.
4. *GS address validation*: AVLC frame has the unique source address and destination address, which can realize the GS recognition.
5. *Retransmission*: The DLS sublayer can retransmit an AVLC frame that is lost or invalid in the destination.

5.2.3.2.3 AVLC Frame Structure

Inherited from the HDLC protocol, AVLC protocol has the similar frame structure. The only difference between them exists in the address field. AVLC extends the address from 8 bits (HDLC frame) to 32 bits (both in source address and in destination address), which can realize the transmitter authentication and enhance the system security (Figure 5.3).

In AVLC frame, the fields are defined as follows:

1. *Flag*: The field shall identify the frame boundary and realize frame synchronization. It is preset to 01111110.
2. *Address*
   a. *Broadcast and multicast*: When the address type is 1/1/1 and the other destination address bits are all 1, the VDL2 works in broadcast mode. When the address type is not 1/1/1 (such as 1/0/0) and other destination address bits are all 1, the VDL2 works in multicast mode.

Flag 8 bits	Address 64 bits	Control 8 bits	Information varied lengths	FCS 16 bits	Flag 8 bits

FIGURE 5.3
AVLC frame structure.

b. *Address type subfield*: 001 represent 24-bit ICAO address space at AC. 100 and 101, respectively, represent the 24-bit ICAO-administered address space and the 24-bit ICAO-delegated address space at GS. Among them, the ICAO-administered address includes a country code prefix (using the same country code assignment defined in the ICAO Annex 10, Volume III, Chapter 9, Appendix 1, table 1) and a suffix, which is assigned by appropriate authority. The ICAO-delegated address is decided by the delegated organization. 111 represent the broadcast address space. The other formats are reserved for future use.

c. *A/G bit*: *The* bit is used to specify the type of the message sender, where 0 represents aircraft (AC) and 1 represents GS.

d. *C/R status bit*: *The* status bit in the source address field shall be the command/response (C/R) bit. The *C/R* bit shall be set to 0 to indicate a command frame and set to 1 to indicate a response frame.

e. *LSB* (*Least Significant Bit*) *bit*: The bit is used to indicate the start and the end of address field and is preset to 00000001.

3. *Control*: The field is the same as the HDLC. According to the different frame types, it can be divided into information (I) frame, supervisory (S) frame, unnumbered (U) frame, etc., control field structure.

4. *FCS*: Frame check sequence (FCS) field, generated by CRC polynomial, supports the error detection function.

5. *Information*: The field is consisted of UD, Initial Protocol Identifier (IPI), and Extended IPI (ExIPI). UD shall bear the data from user with a variable length. IPI and ExIPI can distinguish the message type (CLNP protocol message, ES–IS protocol message, and IS–IS protocol message). According to the message type, packet is delivered to the corresponding module (CLNP, IS–IS, and ES–IS) (Table 5.4).

5.2.3.3 VDL Management Entity

As the highest level of data link layer, VME accepts services from the DLS sublayer and physical layer. At the other side, VME provides services to network layer and can be regarded as the interface between the data link layer and the upper layer.

5.2.3.3.1 Architecture

A VME shall have an LME link for each peer LME. Hence, that is, a ground VME shall have an LME per aircraft, and an aircraft VME shall have an LME per ground system. An LME shall establish a link between a local DLE and a remote DLE associated with its peer LME. A ground LME shall determine if an aircraft station is associated with its peer aircraft LME by comparing the aircraft address; two aircraft stations with identical aircraft addresses are associated with the same LME. An aircraft LME shall determine if a GS is associated with its peer ground LME by bit-wise logical ANDing the DLS address with the GS mask provided by the peer ground LME; two GSs with the identical masked DLS addresses are associated with the same LME.

Each aircraft and ground LME shall monitor all transmissions from its peer's station(s) to maintain a reliable link between some GSs and the aircraft while the aircraft is in the coverage of a GS.

TABLE 5.4

AVLC Frame Bit Structure

Description		Bytes	Bits							
			8	7	6	5	4	3	2	1
Flag		0	0	1	1	1	1	1	1	0
Destination address		1	D22	D23	D24	D25	D26	D27	A/G	0
						Address type				
		2	D15	D16	D17	D18	D19	D20	D21	0
		3	D8	D9	D10	D11	D12	D13	D14	0
		4	D1	D2	D3	D4	D5	D6	D7	0
Source address		5	S22	S23	S24	S25	S26	S27	C/R	0
						Address type				
		6	S15	S16	S17	S18	S19	S20	S21	0
		7	S8	S9	S10	S11	S12	S13	S14	0
		8	S1	S2	S3	S4	S5	S6	S7	1
Link control		9				P/F				
Information	IPI	10	1	1	1	1	1	1	1	1
	ExIPI	11	1	1	1	1	1	1	1	1
	UD	12 ... $N-2$				User data				
FCS		$N-1$								
		N								
Flag		$N+1$	0	1	1	1	1	1	1	0

5.2.3.3.2 Functionality

1. *Frequency management*: When the VDL2 is stated on a given service provider, the VHF digital radio (VDR) is configured to filter uplinks which address matches the provider prefix (10 first bits of the total 24 bits). VME shall scan up to 4 min on CSC, on which GS regularly transmits broadcast frames (GSIF) to announce its presence and position information. Every GSIF during the scan period initiates a link establishment procedure with the uplink address.

2. *GS management*: VME stores all the information relative to the GS in coverage in a peer entity connection (PEC) table and adds signal quality information provided by the VDR with each uplink. VME can use PEC to make connection choice. And if the signal quality of the current connected GS is too low, VME shall initiate a handoff procedure to another GS, which can provide better signal quality.

3. *Link-Change Notifications*: The VME shall notify the intermediate-system system management entity (IS-SME) of changes in link connectivity by supplying information contained in the exchange identity (XID) frames received.

5.2.4 Subnetwork Layer

The subnetwork layer protocol is referred to formally as a subnetwork (across the VHF air–ground subnetwork) access protocol and shall conform to ISO 8208 protocol. On the AC/GS interface, the aircraft subnetwork entity shall act as data terminal equipment (DTE) and the ground subnetwork entity shall act as data circuit-terminating equipment (DCE).

5.2.4.1 Architecture

Data link layer shall strip off the layer 2 header and trailer from the received frame and pass the remaining DLS UD up to subnetwork layer within a DLS primitive. This remainder is a subnetwork protocol data unit (SNPDU).

5.2.4.2 Functionality

1. *Data control*: Subnetwork layer is responsible for controlling data packet flow with respect to duplicate, lost, or invalid SNPDU. Subnetwork layer can break SNPDU into segments, which is called SNPDUs, for data control and error recovery.

2. *Virtual circuit*: Subnetwork layer is basically responsible for internal routing (or relaying) and provides data transferring across the subnetwork. It shall set up and maintain the connection between air and ground routers, which is known as virtual circuit, according to ISO 8208.

5.3 Key Technology

For the VDL2, some technologies will play key roles in bettering performance and have been demanded or recommended by the VDL SARPs. Most of the key technologies are widely used in present communication systems and are well known to the communication engineer or researcher. But the key technologies often have their specialty when used in an aviation air–ground communication system, where media is radio and bandwidth is limited. In the section, several key technologies will be discussed and analyzed under the VDL2 environment.

5.3.1 Physical Layer Key Technology

The principle of physical layer design is to select appropriate technologies according to the real channel characteristics, which can, at system's best, improve the reliability under limited resources (such as power or bandwidth). The choice of technology is decided by channel characteristics, that is, the strategy of physical layer technology is varied with the difference of channels. Essentially, the fundamental purpose of applying the physical layer technology is to enable the system to adapt to the physical channel.

Operating in VHF band, the VDL2 system has good propagation characteristics, which can neglect multipath/selective fading effect and can be considered as having only electromagnetic wave attenuation with the distance. Therefore, there are no anti-fading measures (such as spread spectrum, diversity reception) in the VDL2. In modulation, the VDL2 uses the D8PSK scheme, which can improve the information transmission rate. In channel coding, the VDL2 uses the RS FEC coding, which can enhance the ability of anti-burst error in mobile environment.

5.3.1.1 D8PSK Modulation

In the process of D8PSK implementation, the first step is to implement parallelization (1/3), and next step is to map phase difference based on Gray code. The major characteristic of Gray code is that the distance between adjacent code words is always 1, that is, if

TABLE 5.5

Phase Mapping

X_k	Y_k	Z_k	$\Delta\varphi_k$
0	0	0	$0\pi/4$
0	0	1	$1\pi/4$
0	1	1	$2\pi/4$
0	1	0	$3\pi/4$
1	1	0	$4\pi/4$
1	1	1	$5\pi/4$
1	0	1	$6\pi/4$
1	0	0	$7\pi/4$

a bit error happens in Gray code, the original code word will be replaced by the adjacent one. Obviously, a (Gray code) bit error will cause smaller error in sample receiving than that of natural binary code (Table 5.5).

After the phase-difference mapping, differential encoding needs to be calculated from the previous phase and the phase difference.

$$\Phi_k = \Phi_{k-1} + \Delta\Phi_k \tag{5.2}$$

After the completion of phase encoding, the modulated signal can be obtained by IQ orthogonal modulation.

If the correlation detection is applied for demodulation, the symbol error rate (P_e) under big signal noise ratio (SNR) can be calculated by the following formula[9]:

$$P_e \approx 2Q\left(\sqrt{\frac{2E_s}{N_0}}\sin\frac{\pi}{\sqrt{2}M}\right) \tag{5.3}$$

where

E_s is the energy of a symbol
N_0 is the noise power spectral density
M is the level of coding (for D8PSK, $M=8$)

For phase modulation and Gray code, bit error rate (BER, P_b) can be calculated by the following formula:

$$P_b \approx \frac{P_e}{\log_2 M} \tag{5.4}$$

From (5.3) and (5.4), we can get the BER as per the following equation:

$$P_b \approx \frac{1}{\log_2 M} 2Q\left(\sqrt{\frac{2E_s}{N_0}}\sin\frac{\pi}{\sqrt{2}M}\right) \approx \frac{2}{3}Q\left(\sqrt{\frac{2E_s}{N_0}}\sin\frac{\pi}{8\sqrt{2}}\right) \tag{5.5}$$

where Q function can be derived from error function or complementary error function.

$$Q(x) = \frac{1}{2}erfc\left(\frac{x}{\sqrt{2}}\right) = \frac{1}{2}\left(1 - erf\left(\frac{x}{\sqrt{2}}\right)\right) \tag{5.6}$$

Considering the following equation,

$$\frac{P_r}{N_0} = \frac{E_b}{N_0} R = \frac{E_s}{N_0} R_s \tag{5.7}$$

we have

$$E_b = \frac{R_s}{R} E_s = \frac{R_s}{R_s \log_2 M} E_s = \frac{1}{3} E_s \tag{5.8}$$

where
  P_r is the received power
  E_b is the bit energy
  R is the bit rate
  E_s is the symbol energy
  R_s is the symbol rate

From (5.5) and (5.8), we can get the following equation, which can evaluate the BER performance of D8PSK modulation:

$$P_b \approx \frac{1}{3} 2Q\left(\sqrt{\frac{6E_b}{N_0}} \sin\frac{\pi}{8\sqrt{2}} \right) \tag{5.9}$$

5.3.1.2 RS Channel Encoding

In channel coding, the VDL2 applies RS code. As a q-nary code, RS code is the most important subclass of BCH code. Each element value of RS code is decided by the Q element symbol set:

$$\left\{ 0, \alpha^0, \alpha^1, \ldots \alpha^{q-2} \right\} \tag{5.10}$$

In the process of coding, q value is usually 2 to the power of a positive integer. As a result, the entire nonzero elements (in the Q element set) belong to a mth degree primitive polynomial-based extension field GF (2^m), which represents the Galois field on the finite set of integer 2^m.[10]

In the VDL2 physical layer, channel coding applies RS (255,249) system code in which the code length is fixed and m is 8. The primitive polynomial chosen is as follows:

$$p(x) = (x^8 + x^7 + x^2 + x + 1) \tag{5.11}$$

The generation polynomial can be gotten as follows:

$$g(x) = \prod_{i=120}^{125} (x - \alpha^i) \tag{5.12}$$

where α is the primitive element (in the Galois field GF (2^8)), which is derived from (5.11).

Due to $q = 2^8$ (256-nary), each symbol represents 1 octet and can transmit 8 bits. Therefore, RS (255,249) will make block code for 249 octets (1992 bits) and will add 6 octets (48 bit) redundancy bytes in coding.

The block number k is as follows:

$$k = \frac{\text{Data length (bits)}}{1992 \text{ bits}} \tag{5.13}$$

The error-correcting capability can be calculated from the following formula:

$$t = \text{int}\left[\frac{(d_{min} - 1)}{2}\right] = \text{int}\left[\frac{(n - k)}{2}\right] = \text{int}\left[\frac{r}{2}\right] = \text{int}\left[\frac{48}{2}\right] = 24 \text{ bit} \tag{5.14}$$

The error-correcting ratio (ECC_{ratio}) is

$$ECC_{ratio} = \frac{\text{Error-correctable message}}{\text{Total message}} = \frac{3 \times 8}{255 \times 8} = \frac{1}{85} \tag{5.15}$$

5.3.2 Data Link Layer Key Technology

5.3.2.1 P-CSMA

As a competitive access mode, P-CSMA inherits the essential features of CSMA and increases the probability judgment algorithm.

5.3.2.1.1 Working Mechanism

P-CSMA keeps checking channel status and waits for the channel to be idle. When the channel becomes idle and the transmitting number does not reach the maximum number of access (M_1), the system will transmit queued data with probability p or wait until the next transmission timing with probability $1 - p$. After reaching M_1, the system will transmit immediately under the idle state. If queued data are not transmitted after channel busy timer (TM_2) expiration, the VDL2 MAC sublayer will detect congestion and relay the message to VME. If the channel is busy, the system continuously monitors until the channel becomes idle. After becoming idle, the system will repeat the same as mentioned earlier.

According to whether the time slot is applied, P-CSMA can be classified into two modes[11]:

1. *Slot mode*: Sending time is divided into many small intervals (slots), and media can only be accessed in the front edge of a slot. In this mode, the nodes keep synchronization via the master clock of the network.

2. *Nonslot mode*: Sending time is not divided into slots, and media can be accessed at any time. In this mode, the network synchronization is not needed (Figure 5.4).

5.3.2.1.2 MAC Sublayer Parameters

As the key technology of media access control, P-CSMA is closely relevant to the parameters regulated in the VDL2 MAC sublayer (Table 5.6).

1. *Parameter P*: The parameter p $(0 < p \leq 1)$ is the probability under which the MAC sublayer will transmit on any access attempt that will determine whether the transmitter should be immediately enabled. An access attempt shall be made when

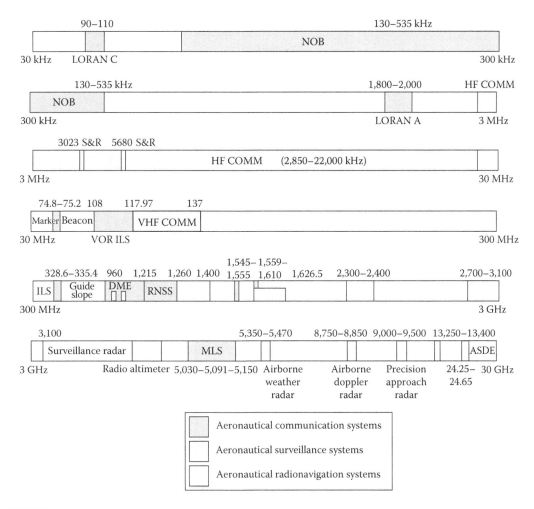

FIGURE 1.4
Communications radio navigation and surveillance bands.

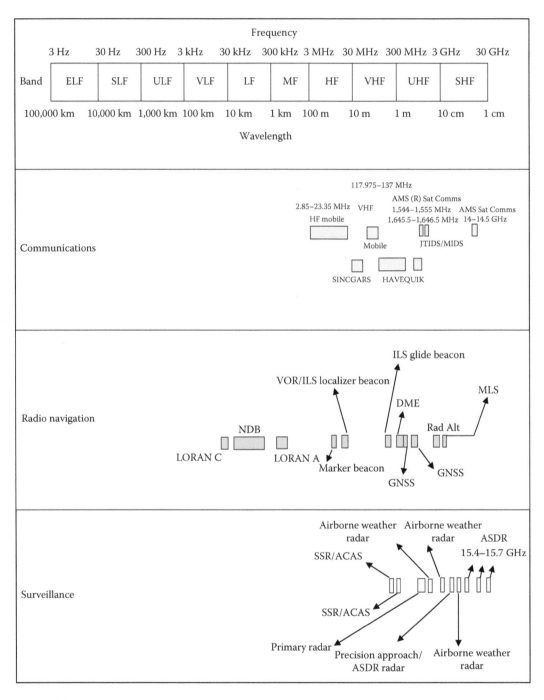

FIGURE 1.5
Aeronautical radio spectrum.

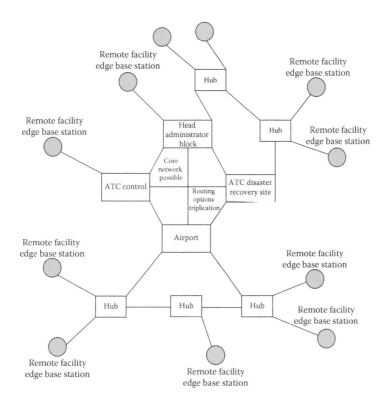

FIGURE 1.6

Hypothetical Aeronautical Telecommunication Network. (From ICAO, *Aeronautical Telecommunication Network (ATN): Manual for the ATN Using IPS Standards and Protocols*, Doc 9896. September 2008.)

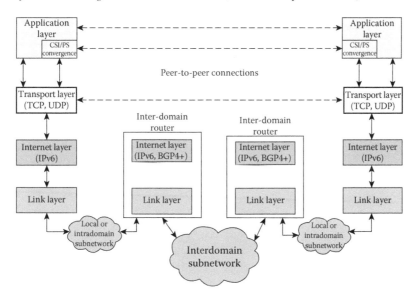

FIGURE 1.7

Aeronautical Telecommunication Network/Internet protocol suite protocol architecture. (From ICAO, *Aeronautical Telecommunication Network (ATN): Manual for the ATN Using IPS Standards and Protocols*, Doc 9896, September 2008.)

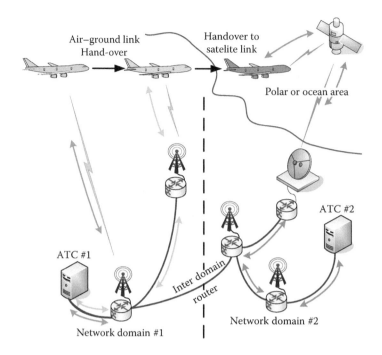

FIGURE 2.1
ATN mobile communication framework.

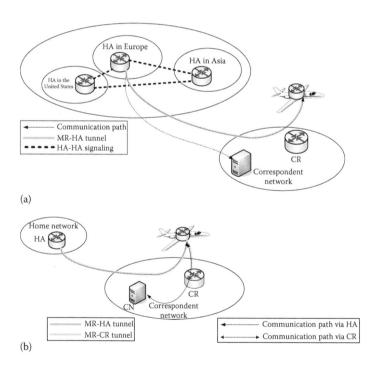

FIGURE 2.10
Typical air traffic services communication scenario making use of different route optimization protocols.
(a) Global HA-to-HA and (b) Communication router (Adapted from Bauer, C., *Secure and Efficient IP Mobility Support for Aeronautical Communications.* Karlsruhe, Germany: KIT Scientific Publishing, 2013.)

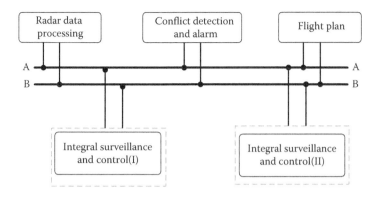

FIGURE 3.1
Composition of ATC.

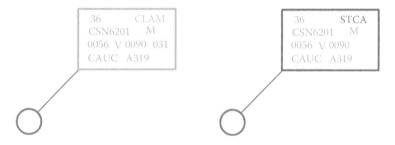

FIGURE 3.4
Conflict warning.

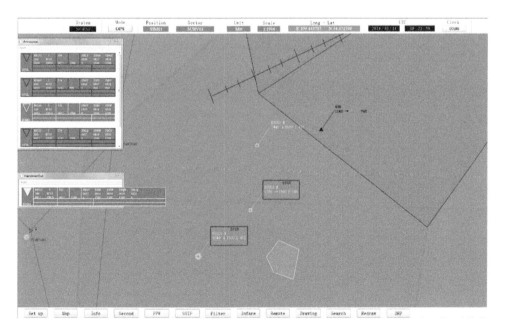

FIGURE 3.7
Overview of ISC.

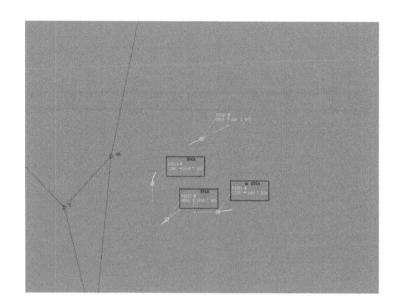

FIGURE 3.8
Track.

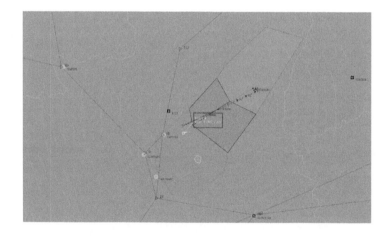

FIGURE 3.9
Maps.

FIGURE 3.10
Flight plan window.

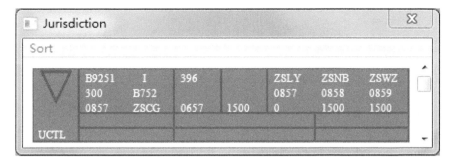

FIGURE 3.11
Flight strip.

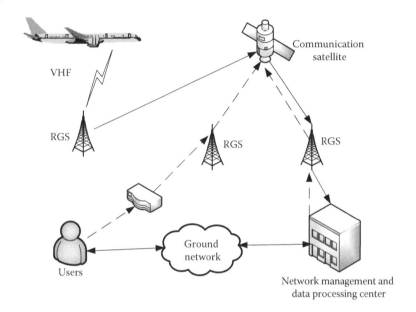

FIGURE 4.1
The composition of ACARS.

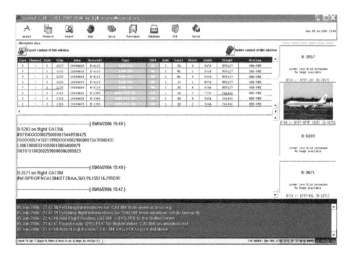

FIGURE 4.2
ACARS receiving software acarsd 1.65.

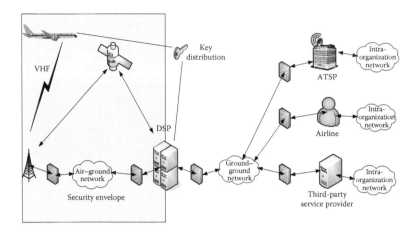

FIGURE 4.3
DSP-based security architecture.

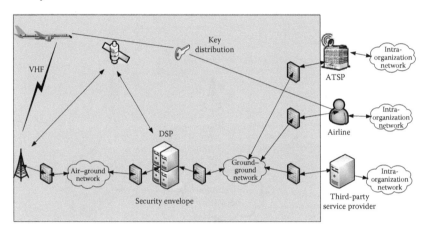

FIGURE 4.4
End-to-end security architecture.

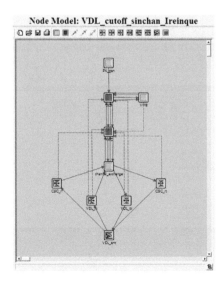

FIGURE 5.7
The GS node model.

Process Model: VDL_GS_csma_cutoff_single_channel_upd

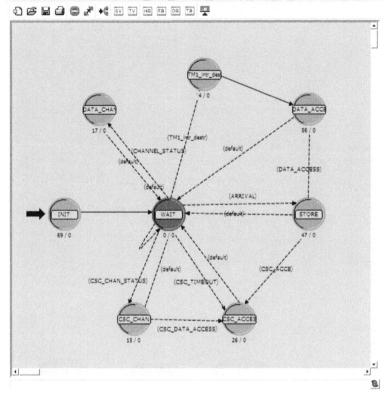

FIGURE 5.9
The GS MAC process.

Process Model: VDL_aircraft_HO_child_upd

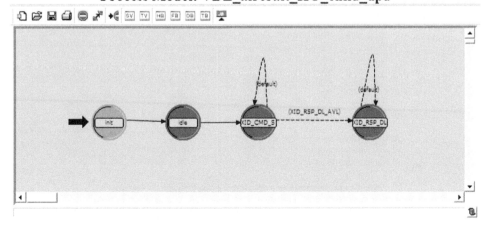

FIGURE 5.10
The AC handoff child process.

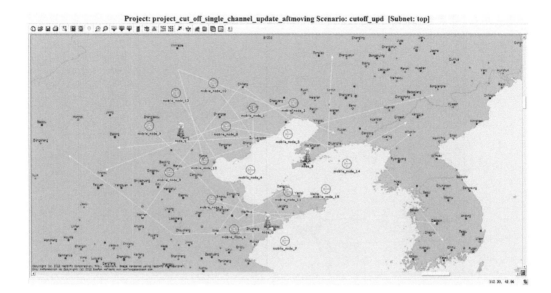

FIGURE 5.14
Handoff scenario.

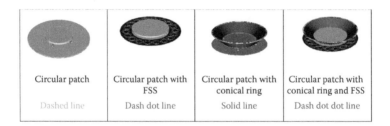

Circular patch	Circular patch with FSS	Circular patch with conical ring	Circular patch with conical ring and FSS
Dashed line	Dash dot line	Solid line	Dash dot dot line

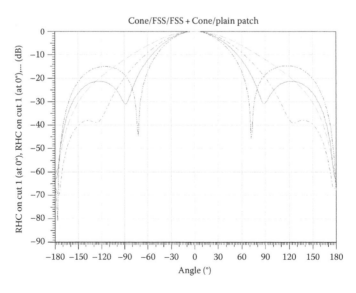

FIGURE 6.7
Patterns of all four structures compared.

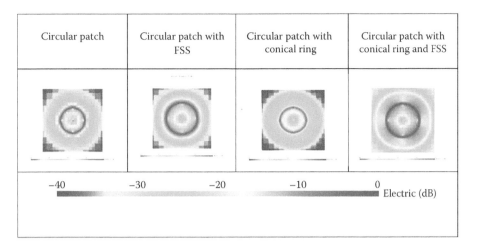

FIGURE 6.8
RMS electric field distributions in a slice taken at 7 mm on top of the ground plane at 1.575 GHz.

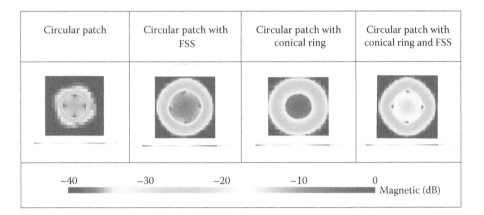

FIGURE 6.9
RMS magnetic field distributions in a slice taken at 7 mm on top of the ground plane at 1.575 GHz.

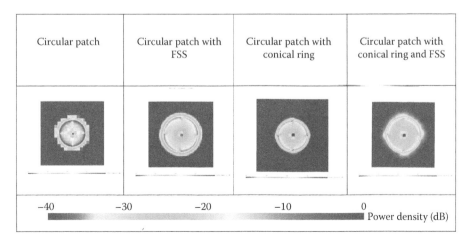

FIGURE 6.10
Time average power density in a slice taken at 7 mm on top of the ground plane at 1.575 GHz.

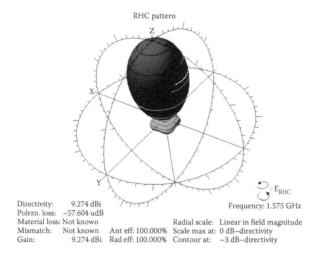

RHC pattern

Directivity:	9.274 dBi			Frequency: 1.575 GHz
Polrzn. loss:	−57.604 udB			
Material loss:	Not known		Radial scale:	Linear in field magnitude
Mismatch:	Not known	Ant eff: 100.000%	Scale max at:	0 dB−directivity
Gain:	9.274 dBi	Rad eff: 100.000%	Contour at:	−3 dB−directivity

FIGURE 6.11
3-D RHC radiation pattern.

Comparing conical ring, FSS, and ordinary circular patch

Directivity:	9.709 dBi			Frequency: 1.575 GHz
Polrzn. loss:	−60.884 dB			
Material loss:	−18.581 mdB		Radial scale:	Linear in field magnitude
Mismatch:	Not known	Ant eff: 99.573%	Scale max at:	0 dB−directivity
Gain:	−51.194 dBi	Rad eff: 99.573%	Contour at:	−3 dB−directivity

FIGURE 6.12
3-D LHC radiation pattern.

FIGURE 6.16
A possible antenna test mount for urban canyon environment.

FIGURE 6.17
A KSU Salina UAS asset. (Aerosonde Mark 4.7 UAS platform.)

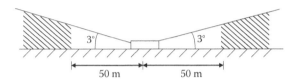

FIGURE 7.3
No obstruction outside 50 m in the bottom center of the antenna as a benchmark over a vertical opening angle of 3°.

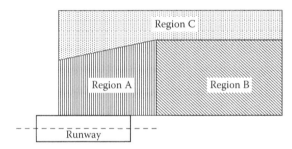

FIGURE 7.8
The protected areas of the glide slope.

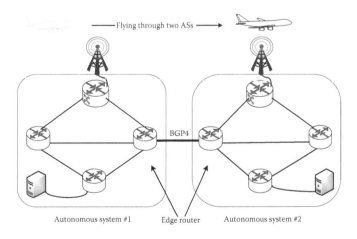

FIGURE 8.6
ATN/IPS routing infrastructure.

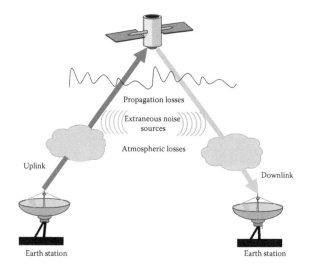

FIGURE 9.2
Satellite link budget.

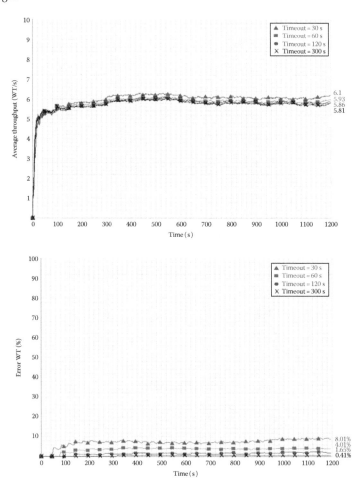

FIGURE 10.7
Four kinds of throughput and timeout probability curve on the condition of trouble-free equipment.

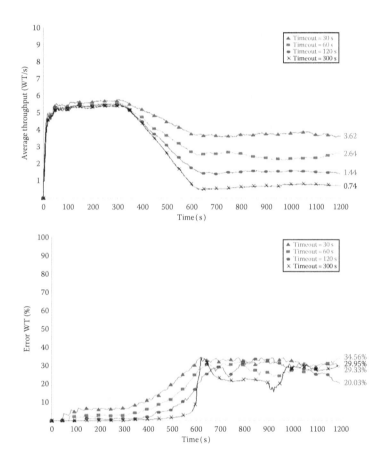

FIGURE 10.8
Variation curves of 10% of equipment breakdown at the 300th second during the experiment.

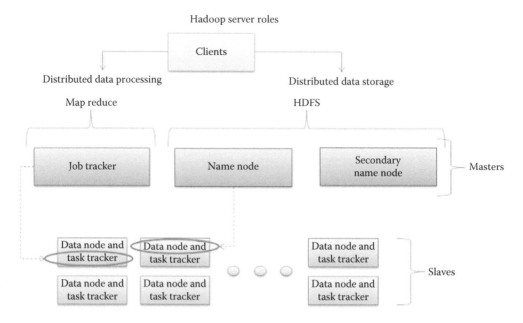

FIGURE 10.12
Hadoop server roles.

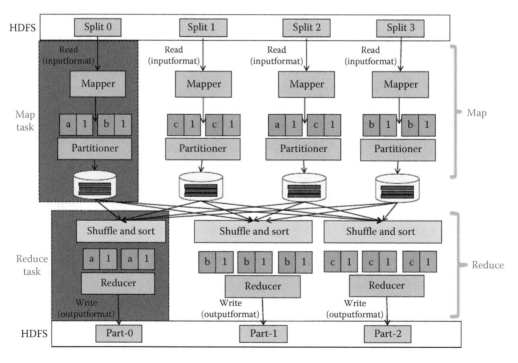

FIGURE 10.13
MapReduce data flow.

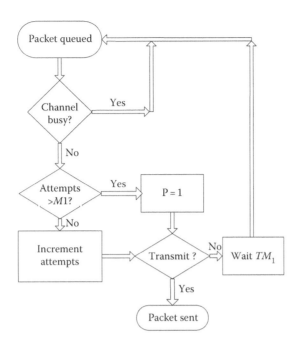

FIGURE 5.4
P-CSMA flowchart.

TABLE 5.6

MAC Sublayer Parameters

	Description	Unit	Minimum	Maximum	Default	Increment
M_1	Maximum number of access	Times	1	65,535	135	1
P	Persistence	No	1/256	1	13/256	1/256
TM_1	Interaccess delay	Millisecond	0.5	125	4.5	0.5
TM_2	Channel busy	Second	6	120	60	1

Timer TM_1 expires and the channel is idle, or when a transmission request arrives from the DLS while the channel is idle, or if the channel is determined to become idle while a message is queued for transmission. The result of an access attempt is decided by the random number produced by P-CSMA: if the random number is less than P, the access attempt is successful; otherwise, the access attempt is unsuccessful. If the access attempt is successful, the transmission shall immediately begin.

2. Counter M_1: The M_1 counter shall evaluate the maximum number of attempts (M_1) that a MAC sublayer will make for any transmission request. This counter shall be started upon system initialization, Timer TM_2 expiring, or a successful access attempt. The counter shall add 1 after every unsuccessful access attempt. When the counter expires, authorization to transmit shall be granted as soon as the channel is idle.

3. Timer TM_1 (interaccess delay timer): The TM_1 timer shall be set to the time (TM_1) that a MAC sublayer will wait between consecutive access attempts. This timer

shall be started if it is not already running, and the channel is idle after an access attempt is unsuccessful. The timer shall be canceled if the channel becomes busy. When the timer expires, another access attempt shall be made.

4. Timer TM_2 (channel busy timer): The TM_2 timer shall be set to the maximum time (TM_2) that a MAC sublayer will wait after receiving a request to transmit. This timer shall be started, if it is not already running, when the MAC sublayer receives a request for transmission. The timer shall be canceled upon a successful access attempt. When the timer expires, the VME shall be informed that the channel is congested.

5.3.2.2 From HDLC to AVLC

AVLC sublayer conforms to HDLC as specified by ISO 13239.[12] However, given that HDLC was designed to primarily support stationary network terminals where bandwidth for the most part is not scarce, AVLC has been optimized to consider the fact that the VDL network terminals are in a mobile environment with limited bandwidth available and to add the adaptive channel estimation algorithm in packet retransmission.

5.3.2.2.1 HDLC Mechanism

In 1974, IBM launched a famous system structure, SNA, in which data link layer protocols use a bit-oriented Synchronous Data Link Control (SDLC). Later, ISO modified SDLC to HDLC as international standard.

Applying serial number and confirmation, HDLC can provide reliable transmission with error free characters, no lost or repeated packets. The HDLC frame is constructed from address field, control field, information field, FCS, and two flags (Figure 5.5).

According to the difference of functionality, the HDLC frame can be divided into three categories, as shown in Table 5.7.

1. *Information frame (I frame)*: I format is used to perform an information transfer.

2. *Supervisory frame (S frame)*: S format is used to perform data link supervisory control functions such as acknowledging I frames, requesting retransmission of I frames, and requesting a temporary suspension of transmission of I frames.

8 bits	8 bits	8 bits	Varible	16 bits	8 bits
Flag (F)	Address (A)	Control (C)	Information (I)	Frame checking sequence (FCS)	Flag (F)

FIGURE 5.5
HDLC frame.

TABLE 5.7

Control Field Structure

	1	2	3	4	5	6	7	8
Information frame (I)	0		$N(S)$		P/F		$N(R)$	
Supervisory frame (S)	1	0		S	P/F		$N(R)$	
Unnumbered frame (U)	1	1		M	P/F		M	

3. *Unnumbered frame* (*U frame*): U format is used to provide additional data link control functions and unnumbered information transfer. This format shall contain no sequence numbers, but shall include a *P/F* bit. Five modifier bit positions are available, thus allowing definition of up to 32 additional command functions and 32 additional response functions.

The functions of *N*(*S*), *N*(*R*), and *P/F* are independent:

1. *N*(*R*) symbols transmitting receive sequence number (bit 6 = low-order bit), which can indicate that the station transmitting the *N*(*R*) has correctly received all I frames numbered up to [*N*(*R*) −1] (inclusive).
2. *N*(*S*) symbols sending sequence number (bit 2 = low-order bit).
3. *P/F is the poll or final bit*: The poll (*P*) bit set to "1" shall be used by the primary/combined station to solicit (poll) a response or sequence of responses from the secondary station(s)/combined station. A response frame with the *F* bit set to "1" shall be used by the secondary/combined station to acknowledge the receipt of a command frame with the *P* bit set to "1."

To reduce the times of acceptance acknowledgment and to improve the transmission efficiency, sliding window and piggybacking are applied in the HDLC procedure:

1. Within the window length (default value is 7), communication peers can continuously send I frame, which is not acknowledged by the peer.
2. Acknowledgment can be piggybacked by I frame from the peer, and the frames from 0 to [*N*(*R*) −1] (mode 8) can be confirmed by *N*(*R*) bit at once.

To avoid the transmitter waiting acknowledgment for an undesirable lag, HDLC introduces the time-out retransmission: whenever the transmitter sends an I frame, it will start the retransmission timer and will keep timing until receiving confirmation (including piggybacking) from the peer. When a frame is correctly received, the receiver will initiate its timing process. The timer will be canceled when the receiver has sent at least a frame that piggybacks the acknowledgment information. At the expiration of the receiver timer, an S frame will be sent to inform the transmitter that the transmitted frames has been received correctly.

5.3.2.2.2 AVLC Improvement on HDLC

Originating from HDLC, AVLC has the inheritance with HDLC and can be regarded as a subset of HDLC. It applies the adaptive channel estimation algorithm in the HDLC retransmission, which is the most different point from HDLC and the most prominent improvement on HDLC. The algorithm is proposed to adapt the characters of terminal mobility and frequency scarcity existing in the air–ground data link control.

HDLC uses a constant timing retransmission in which there exists its shortcoming: on one hand, when the channel utilization rate is comparably high, retransmission frame after a constant delay before retransmission (T_1) is easy to collide in P-CSMA and will further aggravate the channel burden. On the other hand, when the channel utilization rate is comparably low, retransmission frame after a constant T_1 will lead to unnecessary delay. The shortcomings become obvious and unbearable in a frequency-scarce system.

TABLE 5.8

T_1 Parameters

Symbol	Parameter Name		Lower Bound	Upper Bound	Default	Increment
T_{1min}	Delay before	Minimum	0 s	20 s	1.0 s	1 ms
T_{1max}	retransmission	Maximum	1 s	20 s	15 s	1 ms
T_{1mult}		Multiplier	1	2.5	1.45	0.01
T_{1exp}		Multiplier	1	2.5	1.7	0.01

To overcome the shortcomings, AVLC applies the adaptive channel estimation algorithm for T_1 timer. It is based on the idea that radio channel has a continuous character and the future channel utilization rate can be predicted from that of the current. In the algorithm, the predicted value is applied to adjust T_1 by increasing it with a smaller channel utilization rate and decreasing it with a bigger channel utilization rate.

The T_1 adaptive channel estimation algorithm is as follows:

$$T_1 = T_{1\min} + 2 \times TD_{99} + \min\left(u(x), T_{1\max}\right) \tag{5.16}$$

where

$$TD_{99} = \frac{TM_1 \times M_1}{1 - \mu} \tag{5.17}$$

In (5.17), TD_{99} is the running estimate for the 99th percentile transmission delay (between the time at which the frame is sent to the MAC sublayer and the time at which its transmission is completed). Parameter μ is a measurement of channel utilization with a range of value from 0 to 1, where 1 corresponding to a channel that is 100% occupied. TM_1 and M_1 are parameters of P-CSMA.

The parameter $u(x)$ is a uniform random number generated between 0 and x that is calculated from the following formula:

$$x = T_{1mult} \times TD_{99} \times T_{1exp}^{retrans} \tag{5.18}$$

In (5.18), the superscript ^{retrans} is the largest retransmission count of all the outstanding frames.

T_{1mult}, T_{1exp}, T_{1min}, and T_{1max} are configurable parameters, which are listed in Table 5.8.

From the earlier algorithm, we can see that with the channel utilization (μ) increasing, transmission delay (TD_{99}) and x will increase. With the increase of x, $u(x)$ will increase in the probability of uniform distribution. Moreover, retransmission (T_1) will increase with the probability growing up. As a result, T_1 and μ will change in the same direction, that is, retransmitted frames will be sent slower when channel utilization rate is higher; retransmitted frames will be sent faster when channel utilization rate is lower.

5.3.2.3 Handoff

An aircraft flies along its flight path, and it will move into or out of range of various VDL GSs, which will lead to the change of signal quality. To provide continuous data link communications, the aircraft radio must switch its linking from the current GS to a new GS, which can supply a better signal. The process is called handoff.

The VDL2 supports four variations of the handoff, including aircraft-initiated handoff, aircraft-requested and ground-initiated handoff, ground-initiated handoff, and ground-requested and aircraft-initiated handoff. Aircraft-initiated handoff should be the usual mechanism for a light to moderately loaded subnetwork. Ground-initiated handoff and ground-requested and aircraft-initiated handoff occur under heavy loading conditions to perform load balancing between GSs. A ground-initiated handoff must be to another GS on the same frequency. Therefore, the ground-requested and aircraft-initiated handoff is included to provide for frequency changes. An aircraft-requested and ground-initiated handoff is provided to allow an aircraft to decide when a handoff should occur but allows the ground network to decide which is the best GS for load balancing.[13]

5.3.2.3.1 Aircraft-Initiated Handoff

Once the aircraft LME has established a link to a GS, it shall monitor the VHF signal quality on the link and the transmissions of the other GSs. The aircraft LME shall establish a link to a new GS due to the following reasons:

1. *Signal quality*: The VHF signal quality on the current link is poor, and the signal quality of another GS is significantly better.
2. *Link failure*
   a. The counter N_2 is exceeded on any frame sent to the current GS.
   b. The timer TG_2 expires for the current link.
   c. The timer TM_2 expires.

In the way of aircraft-initiated handoff, an aircraft decides when to start initiating handoff and sends an XID_CMD_HO message to the newly chosen GS in order to request the handoff. If the request is accepted, the GS will respond with an XID_RSP_HO message; otherwise, the GS will reject the proposed handoff with an XID_RSP_LCR message. When a handoff has been accepted, the aircraft terminates the old data link (1) by transmitting a DISC frame, perhaps after waiting for a short time to ensure that all data have been transferred over the old data link. This procedure is performed when handoff occurs between service providers or (2) by both avionics and the old GS starting a timer (TG5) and automatically clearing the calls at timer expiry. This procedure is used when handoff occurs between GSs belonging to the same service provider and implies that some sort of message is sent between new and old GSs (Figure 5.6).

5.3.2.3.2 Ground-Initiated Handoff

In the way of ground-initiated handoff, an aircraft decides when to start initiating handoff and sends an XID_CMD_HO message to an aircraft in order to request the handoff. If the request is accepted, the aircraft will respond with an XID_RSP_HO message. The handoff decision and TG5 timing of this mode are similar to aircraft-initiated handoffs and will not be described here.

5.3.2.3.3 Aircraft-Requested and Ground-Initiated Handoff

In the way of aircraft-requested and ground-initiated handoff, an aircraft first sends an XID_CMD_HO message to a GS and does not require immediate response. The GS will hand over the message, via ground–ground communication, to the next GS, which will

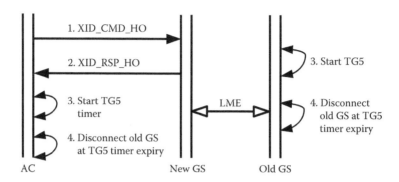

FIGURE 5.6
The process of aircraft-initiated handoff.

decide when to start initiating handoff and will send an XID_CMD_HO ($P=1$) message to the aircraft. If the request is accepted, the aircraft will respond with an XID_RSP_HO ($F=1$) message. The handoff decision and TG5 timing of this mode are similar to aircraft-initiated handoffs and will not be described here.

5.3.2.3.4 Ground-Requested and Aircraft-Initiated Handoff

In the way of ground-requested and aircraft-initiated handoff, a GS first sends an XID_CMD_HO message to an aircraft and does not require immediate response. The aircraft will decide which GS to select and when to start initiating handoff. Then, it will send an XID_CMD_HO ($P=1$) message to the selected GS. If the request is accepted, the GS will respond with an XID_RSP_HO ($F=1$) message. The handoff decision and TG5 timing of this mode are similar to aircraft-initiated handoffs and will not be described here.

5.3.3 Subnetwork Link and Connection

The VDL2 subnetwork layer is an implementation of ITU-T X.25 as specified in ISO 8208 and can supply a variety of facilities to establish, terminate, and manage connections.

5.3.3.1 Explicit Subnetwork Connection

The connection of an explicit subnetwork is based on the prerequisite of link establishment, which is implemented by LME. If no link remains, an aircraft transmitting/receiving a DM frame shall initiate link establishment with a GS. The aircraft LME shall choose a GS, with which it wishes to establish a link based on the signal quality of all received uplink frames and on information in any received GSIFs, and then shall attempt to establish a link with the chosen GS by sending an XID_CMD_LE ($P=1$) message. If the ground LME receives the message, it shall confirm the link establishment by sending an XID_RSP_LE ($F=1$) message and complete the link procedure.

During link establishment, a ground DCE shall indicate its available routers in the ATN outer NETs parameter, and the aircraft LME shall then attempt to maintain all subnetwork connections.

1. *Explicit subnetwork connection establishment*: Immediately after link establishment, the aircraft DTE shall attempt to establish a subnetwork connection to at least

one ground DTE. The aircraft DTE shall request a single subnetwork connection per ground DTE by the transmission of a CALL REQUEST packet specifying the ground DTE address. On receipt of the CALL REQUEST, the ground DCE shall attempt to establish a subnetwork connection to the aircraft DTE by responding with a CALL CONFIRMATION packet; otherwise, the ground DCE shall send a CLEAR REQUEST packet including the clearing cause and diagnostic code of the failure. If Ground Network X.121 DTE addressing is implemented, the ground DCE shall use the Called Line Address Modification Notification facility to inform the aircraft DTE of the ground DTE's X.121 address. Else, if the default ground DTE addressing is implemented, the ground DCE shall use the Called Address Extension facility to inform the aircraft of the ground DTE's Ground VDL-Specific DTE Addressing (VSDA) address that was delivered in the CALL REQUEST.

2. *Explicit subnetwork connection maintenance*: To explicitly request subnetwork connection maintenance to a ground DTE, an aircraft DTE shall send a CALL REQUEST packet to the ground DTE with the fast select facility set containing a VDL mobile SNDCF Call UD Field indicating a request to maintain SNDCF context. If the ground DTE can accept the call, it shall respond with a CALL CONFIRMATION packet with the fast select facility set containing a VDL mobile SNDCF Call UD field indicating whether the SNDCF context was maintained. If the ground DTE or a DCE is unable to accept the call, it shall send a CLEAR REQUEST packet to the aircraft DTE including the clearing cause and diagnostic code of failure. If Ground Network X.121 DTE addressing is implemented, then the ground DTE shall use the Called Line Address Modification Notification facility to inform the aircraft DTE of the ground DTE's X.121 address. Else if the default ground DTE addressing is implemented, then the ground DCE shall use the Called Address Extension facility to inform the aircraft of the ground's VSDA address that was delivered in the CALL REQUEST.

5.3.3.2 Expedited Subnetwork Connection

To perform an expedited subnetwork connection establishment or maintenance, the initiating LME shall include in the XID_CMD the Expedited Network Connection parameter for each subnetwork connection that needs to be established or maintained. If the responding LME receives an XID_CMD with one or more Expedited Network Connection parameters, it shall confirm subnetwork connection establishment or maintenance by sending an XID_RSP containing the parameters. This function shall be applicable only for the link establishment, air-initiated handoff, and ground-initiated handoff processes.

1. *Expedited subnetwork connection establishment*: The aircraft DTE shall reissue CALL REQUESTs for those logical channels for which responses (i.e., either a CALL CONFIRMATION or a CLEAR REQUEST) were not included in the XID_RSP_LE. If Ground Network X.121 DTE addressing is implemented, then the ground DCE shall use the Called Line Address Modification Notification facility to inform the aircraft DTE of the ground DTE's X.121 address. Else, if the default ground DTE addressing is implemented, the ground DCE shall use the Called Address Extension facility to inform the aircraft of the ground's VSDA address that was delivered in the CALL REQUEST.

2. *Expedited subnetwork connection maintenance*: The initiating DTE shall reissue CALL REQUESTs for those logical channels for which responses (i.e., a CALL CONFIRMATION or a CLEAR REQUEST) were not included in the XID_RSP_HO. A ground DTE shall include its Calling Address in the appropriate field. If Ground Network DTE addressing is implemented, then the ground DTE shall use the Called Line Address Modification Notification facility to inform the aircraft DTE of the ground DTE's X.121 address. Else, if the default ground DTE addressing is implemented, the ground DCE shall use the Called Address Extension facility to inform the aircraft of the ground's VSDA address that was delivered in the CALL REQUEST.

5.4 Modeling and Simulation

In this section, a VDL2 model, which includes three-layer protocol stack and aircraft/GS nodes, is established on OPNET platform. Based on the established model, simulations are carried out to analyze and optimize the performance of the data link.

5.4.1 VDL2 Modeling

Aircraft node and GS node are two kinds of nodes existing in the VDL air–ground data link. They are the peers in the half-duplex communication, in which only one can transmit message to the other at any time. They also have a similar protocol structure. As shown in Figure 5.7, the node model can be divided into three layers from the top to the bottom corresponding to the VDL2 protocol stack: (1) packet layer (subnetwork layer) constructed

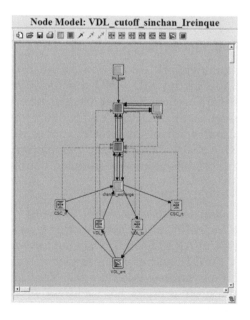

FIGURE 5.7
(See color insert.) The GS node model.

from PK_gen module, (2) data link layer constructed from DLE/VME and MAC modules, and (3) physical layer constructed from PK_gen module, which can implement the switch between transmitter and receiver.

In modeling, the VDL2 channel can be divided into message channel, which is responsible for point-to-point communication, and CSC channel, which is responsible for broadcasting in the band within 136.975–137 MHz. A transmitter/receiver module will occupy only a message or CSC channel and will be set up according to the channel type. Also, a peer of transmitter and receiver work in the same frequency within a 25 kHz band. On the bottom of aircraft node model is the omnidirectional antenna, which can communicate with transmitter and receiver in the way of packet stream.

The difference between AC and GS nodes mainly lies in that aircraft can implement handoff between multiple GSs. Accordingly, the aircraft side adopts a multiple message-transceiver scheme, which is different from the GS side with the single message transceiver.

5.4.1.1 Physical Link Layer Modeling

The physical link layer is modeled in PK_gen module and transmitter/receiver module and radio pipeline, implementing modulation, channel encoding, and physical frame framing and switching.

1. *D8PSK modulation*: In OPNET, modulation scheme is realized in a modulation curve, which can reflect the relationship of BER and E_b/N_0. Therefore, we realized D8PSK modulation by Equation 5.9 using the modulation curve editor.

2. *RS encoding*: The RS encoding error-correcting ratio can be calculated in Equation 5.15 and is set as the receiving SNR threshold in the error correction pipeline stage. The key code is as follows:

```
/* Test if bit errors exceed threshold. */
if (pklen = = 0)
  accept = OPC_FALSE;
else
  accept = (((double) num_errs)/pklen) < = ecc_thresh) ? OPC_TRUE:
    OPC_FALSE;
/* Set flag indicating accept/reject in transmission data block. */
op_td_set_int (pkptr, OPC_TDA_RA_PK_ACCEPT, accept);
```

3. *Physical frame framing*: In the VDL2, the data link layer frame is called the AVLC frame, and the physical layer frame is called the physical frame. From top to bottom, the AVLC frame will be encoding, header-adding, and packaging, and finally be converted into the physical frame. The physical frame is the basic data unit to transmit in the physical media and carries the node location, power, conflict, etc., information.

   The frame can be formed in the frame editor. The physical frame model is as given in Figure 5.8.

4. *Physical packet switching*: For a sending packet, the task of physical frame switching is to realize the mapping from sending address to transmitter. For a received packet, the task is to distinguish whether the destination address (of the received packet) is consistent with the node address: if it is inconsistent, the received packet will be destroyed; if it is consistent, the mapping from the address to data flow will be implemented.

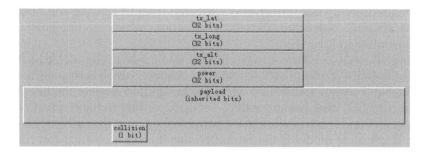

FIGURE 5.8
Physical frame model.

The mapping relationship is achieved through the packet stream ID. The key code is as follows:

```
/* Packet communication for frame being sent in aircraft side. */
if(op_intrpt_strm() = =4)
    op_pk_send(radio_pk,0);
if(op_intrpt_strm() = =5)
    op_pk_send(radio_pk,2);
if(op_intrpt_strm() = =6)
    op_pk_send(radio_pk,4);
```

To facilitate switching modeling, one GS is allocated only one 25 kHz channel, and the channels are differed between different GSs. By this way, the transceivers of GS and aircraft compose a one-to-one mapping, which makes the switching simplified.

5.4.1.2 Data Link Layer Modeling

5.4.1.2.1 MAC Module

The key of MAC process modeling is to realize P-CSMA protocol, including TM_1 timer and P probability decision.[14]

1. *GS MAC modeling*: As shown in Figure 5.9, state INIT fulfills the tasks of variable initialization, object parameters acquirement, statistic registration, and TM_1 timer initialization. DATA_CHAN reads Channel busy/idle statistics from physical layer and updates message to the media access. State *TM1_intr_des* implements TM_1 timer.

   State STORE realizes data stream shunting function with the application of queue. The received data stream is transmitted to the upper layer in a transparent way. According to the type of information, data packets are delivered into different queues waiting for processing. DATA_ACCE implements the algorithm of P-CSMA and realizes the probability decision as follows:

   ```
   /* P probability decision */
   p=op_dist_uniform(1.);
   if((channel_status = =IDLE)&&(p< = csma_p||counter_m1 > = csma_nb_max))
   {Message is dealt and sent.}
   ```

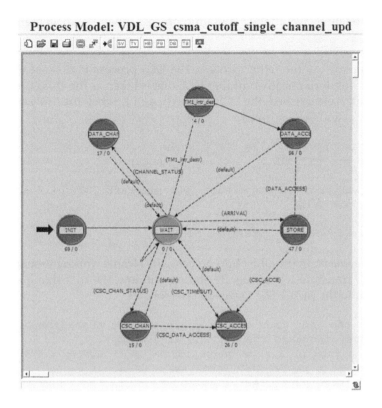

FIGURE 5.9
(See color insert.) The GS MAC process.

The State type of CSC is in symmetry with that of DATA and realizes processing and transmitting of signaling in CSC channel.

2. *AC MAC modeling*: The modeling method of AC MAC is similar to that of GS MAC. The main difference exists in that not only one GS can be connected and switched to an AC. As a result, there are multiple channels in an AC model.

In realizing the conversion of multiple frequencies, AC modeling applies the mechanism of dynamic process. According to the current connection, the parent process dynamically creates and destroys the child process, which is corresponding to the channel with a fixed frequency. In MAC model of AC side, the parent process is responsible for channel assignment, process invoking, and CSC signaling processing. The child process is responsible for implementation of the functions of a fixed channel such as UD sending and probability decision.

5.4.1.2.2 DLS Module

The key of the DLS process modeling is to realize AVLC protocol. In the DLS model, the GS side and aircraft side are all constructed with multiple DLE, single LLC, and sending queue. A DLE corresponds to a logical link between GS and AC, which implements flow control and deals with I frame and S frame. LLC is responsible for the control of logical link and deals with U frame.

In the DLS modeling, we applied dynamic process, queue management, etc., OPNET mechanisms.[15]

1. *Dynamic process*: As the father process, the DLS process is first set up. It will be responsible for dynamically building DLE that is taken as the child process. When a logical link is established, the DLS will birth a DLE; when the link is released, the child process will be destroyed.

```
//create and invoke child DLE_process.
    if(op_pro_valid(DLE_proc)==OPC_FALSE)
        {
        DLE_proc=op_pro_create("VDL_aircraft_DLE_child_upd",
          OPC_NIL);
        }
    op_pro_invoke(DLE_proc,OPC_NIL);
```

2. *Queue management*: In the DLE child process, the I frame waiting to send is inserted into a FIFO transmission queue. By sliding window, the first frame in the queue will be sent to the MAC. The key code is as follows.

```
/* Remove the head from the subqueue and set the NS/NR bits. At the
   same time, increase the current window code. */
s_pkptr=op_subq_pk_remove(queue_id,OPC_QPOS_HEAD);
op_pk_nfd_set(s_pkptr,"NS",local_ns);
op_pk_nfd_set(s_pkptr,"NR",local_nr);
m++;
```

5.4.1.2.3 VME Module

In the VME module, we realized handoff functionality based on dynamic signaling, decision timing, mobile node, multifrequency access, etc., methods.

1. *Dynamic signaling*: From Section 5.3.2.3, we know that handoff will be implemented via signaling interaction between AC and GS. We realized the process by finite-state machine (FSM), using a dynamic process that can simplify the FSM structure and improve the simulation efficiency.

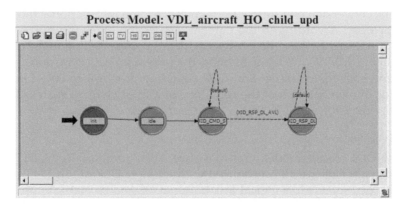

FIGURE 5.10
(See color insert.) The AC handoff child process.

TABLE 5.9

Frequency Configuration

GS	Lower Limit (MHz)	Upper Limit (MHz)	Band (kHz)
node_0	136.000	136.025	25
node_1	136.325	136.350	25
node_2	136.650	136.675	25

Dynamic signaling interaction asks for a strict logical relationship: only after having received the signaling from peer, the local VME can send its own signaling out. In order to guarantee the correct logic, we used a method of stream interrupt combined with unforced state. As in Figure 5.10, the AC child process first sends the XID_CND_HO command and then returns to unforced state. It will wait for response from GS and enter into sleep. If the GS accepts the handoff request and sends the command response, the response will be via packet stream sent to the parent process and trigger the stream interruption. The parent process will wake up the child process. The child process will make further processing and then transfer to the next signaling state.

2. *Decision timing*: The decision of handoff requires a TG5 timer, which is realized via self interrupt in VME module. The key code is as follows:

```
//define transition from the unforced state Idle to the forced state
  HO_destroy.
#define HO_destr op_intrpt_type()==OPC_INTRPT_SELF&&op_intrpt_
  code()==TG5_int
//init_TG5 is the default value of TG5 in VDL2 protocol.
#define init_TG5 20
op_intrpt_schedule_self(op_sim_time()+init_TG5, TG5_int);
```

3. *Mobile node*: Different from the fixed node GS, the AC node is mobile with high speed. The mobility is closely related to the handoff process and is realized via trajectory with segment-based variable interval in modeling.

4. *Multifrequency access*: In the handoff process, AC will possibly switch between GSs with different frequencies. To simplify the simulation, one GS is allocated one single frequency that is far away from the others as possible to weaken the effect of channel cross talk in simulation. The frequency configuration is implemented in the GS antenna module as in Table 5.9.

5.4.1.3 Packet Layer Modeling

Since the VDL2 simulation focuses on the performance of the air–ground data link, the ATN higher level and interface are not involved. Therefore, we take only the packet layer as a data source, that is, a traffic model, ignoring the ATN concerned function. The traffic model can provide data flow to the lower layer and is the basic means of verification and simulation of the network. The design of traffic model should be scientific to ensure a reasonable and reliable simulation.

In the VDL2, the traffic has burst characteristics in packet strength and random characteristic in packet length. Accordingly, we abstracted the traffic model with two stochastic processes, respectively: (1) packet strength modeled with Poisson distribution and (2) packet length modeled with uniform distribution.

5.4.1.3.1 Stochastic Processes

5.4.1.3.1.1 Packet Strength

Definition 5.1: Let X_T be a random process. For any $n(n \geq 2)$ and $t_i(t_0 < t_1 \cdots < t_n)$, if the incremental $X(t_1) - X(t_0)$, $X(t_2) - X(t_1)$, ... , $X(t_n) - X(t_{n-1})$ are independent of each other, then X_T is called an independent increment process.

Definition 5.2: Let X_T be a random process, $X_T = \{N(t), t \in T, T \in [0, \infty)\}$. If $N(t)$ is a nonnegative integer and $N(s) < N(t)$ when $s < t$, then X_T is called a counting process.

Definition 5.3: If the counting process $X_T = \{N(t), t \geq 0\}$ satisfies the following three conditions:

1. $N(0) = 0$
2. X_T is an independent increment process
3. For any nonnegative s and t, there is the equation

$$P\{N(s+t) - N(s) = k\} = e^{-\lambda t} \frac{(\lambda t)^k}{k!} \quad k = 0, 1, 2, \ldots; \lambda > 0 \tag{5.19}$$

then, X_T is called a Poisson process with Poisson strength λ.[16]

With the burst characteristics, packet birth can meet the following conditions:

1. In the nonoverlapping period, the arrivals of packets are independent.
2. The probability of packet arrival is only proportional to Δt and has nothing to do with the starting point.
3. At any small enough period, the number of arrival packets is not more than one; otherwise, it will be considered as impossible event.

With these conditions, we can draw a conclusion that the birth of burst packets meets the Poisson process.

Proof:

1. From the three conditions of burst packets, it is easy to know that the birth of burst packets is both an independent increment process and a counting process.
2. For any Δt, we can know that the arrival probability of a packet is Δt from condition (2) and the nonarrival probability of a packet is $1 - \Delta t$ from condition (3). Assuming that k packets are birthed during period t, we can divide t into n little Δt, which is small enough and contains not more than one packet.

Let

$$P_k(s,t) = P\{N(s+t) - N(s) = k\}$$

we have

$$P_k(s,t) = \lim_{n \to \infty} C_n^k \left(\lambda \Delta t\right)^k \left(1 - \lambda \Delta t\right)^{n-k} = \lim_{n \to \infty} \frac{n!}{k!(n-k)!} \left(\lambda \Delta t\right)^k \left(1 - \lambda \Delta t\right)^{n-k}$$

$$= \lim_{n \to \infty} \frac{n!}{k!(n-k)!} \left(\lambda \frac{t}{n}\right)^k \left(1 - \lambda \frac{t}{n}\right)^{n-k} = \frac{(\lambda t)^k}{k!} \lim_{n \to \infty} \frac{n!}{(n-k)!} \left(\frac{1}{n}\right)^k \left(1 - \lambda \frac{t}{n}\right)^{n-k}$$

$$P_k(s,t) = \frac{(\lambda t)^k}{k!} \lim_{n \to \infty} \frac{n!}{n^n} \frac{(n-\lambda t)^{n-k}}{(n-k)!} \tag{5.20}$$

Due to $n \to \infty$, $n - k \to \infty$, we have

$$\lim_{n \to \infty} \frac{n!}{n^n} \frac{(n-\lambda t)^{n-k}}{(n-k)!} = \lim_{n \to \infty} \left(1 - \frac{\lambda t}{n}\right)^n = \lim_{n \to \infty} e^{\ln\left(1 - \frac{\lambda t}{n}\right)^n} = e^{-\lambda t} \tag{5.21}$$

From Equations 5.20 and 5.21,

$$P_k(s,t) = \frac{(\lambda t)^k}{k!} e^{-\lambda t} \tag{5.22}$$

5.4.1.3.1 Packet Length   In the VDL2, the AVLC packet length l is random and can be presumed to meet uniform distribution.

$$f(l) = \begin{cases} \dfrac{1}{b-a} & a \le l < b \\ 0 & l < a \text{ or } l \ge b \end{cases} \tag{5.23}$$

5.4.1.3.2 Traffic Model

From random distribution theory, we know that if the packet arrival interval obeys exponential distribution with the mean $1/\lambda$, the packet sending process is a Poisson process with strength λ and vice versa. According to the earlier analyses, we made the traffic model flowchart as follows (Figure 5.11):

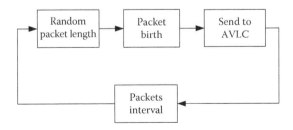

FIGURE 5.11
Traffic model flowchart.

In modeling, random functions are called

```
/* the function to create uniform distribution */
op_dist_uniform(msg_max_length-msg_min_length)
/* the function to create exponential distribution */
op_dist_exponential(iat)
```

Also, self interrupt is applied to trigger the next packet birth and is set to the sum of current moment and exponential distribution-based interval:

```
next_pk_time=op_sim_time()+op_dist_exponential(iat);
op_intrpt_schedule_self(next_pk_time,NEW_PK_TIME);
```

5.4.2 Handoff Simulation

Based on the VDL2 model, simulations are carried out on throughput and delay. The results are similar to the current research. For example, given that payload is defined as the ratio of throughput and system transition rate (31,500 bits/s), we made the throughput versus payload simulation on the model. The result (Figure 5.12) is in consistency with Bretmersky and Kerczewski's experimental result.[17]

In the section, we will focus on the VDL2 handoff and propose a handoff decision method based on interference plus noise ratio (SINR). Simulations will be carried out to comprehend the performance of the method.[18]

5.4.2.1 SINR-Based Handoff

From Section 5.3.2.3, we know that there exist four handoff mechanisms in the VDL2: aircraft-initiated handoff is for the moderately loaded subnetwork; ground-initiated and ground-requested aircraft-initiated handoffs occur under heavy loading conditions;

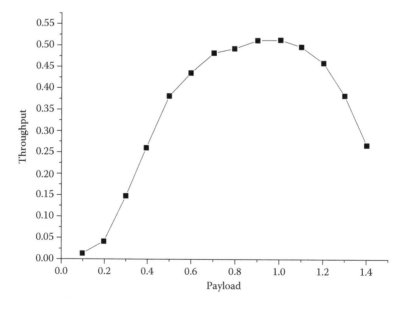

FIGURE 5.12
Throughput versus payload.

aircraft-requested ground-initiated handoff provides ground decision for load balancing. However, how to choose an appropriate mode is a complex and difficult task under a specified condition. It requires a rational and practical switching algorithm that has switching overhead itself. In fact, aircraft-initiated handoff is the usual mechanism in the current VDL2 handoff application.

We try to find a way to optimize aircraft-initiated handoff with the goal of (1) keeping reliable and stable signal strength, (2) balancing the subnetwork load between GSs in a continuous way, (3) considering the handoff overhead, and (4) using a simple algorithm.

5.4.2.1.1 SINR Criteria

As we know, signal attenuation is inversely proportional to the n square of propagation distance, in which n depends on the transmission medium. Due to the distance effect, the signal strengths will change inversely when an aircraft flies from one GS to another GS. After the changing arrives at a certain threshold, a handoff has to be initiated from the original GS to a new GS to keep a good signal quality. Here, signal strengths can be described with SNR, due to the fact that the noise in the VDL2 channel can be regarded as Gauss white noise, and the frequency band is fixed.

However, there exists another effect in the VDL2 network: the distribution of high-speed nodes will change in a constant way, which will lead to the nodes density uneven. For the competition-based P-CSMA, high density will cause an increase in media access delay. Besides, uneven density will cause some channels to be too busy as well as some channels to be too idle and result in an overall system capacity decline.

However, in P-CSMA, multiple nodes will access the channel in a preemptive way. The media access process can be regarded as co-frequency interference. Due to packet birth being a Poisson process, we can deduce that the signal interference ratio (SIR) is in negative correlation with node density. Therefore, we can describe node density with SIR. Furthermore, we take SINR as the handoff criterion to eliminate the negative effects of uneven density.

Definition 5.4: SINR is the ratio of received signal strength to the sum of Gauss white noise and co-frequency interference.

From the SINR definition, we have

$$SINR = \frac{1}{\dfrac{1}{SNR} + \dfrac{1}{SIR}} \tag{5.24}$$

5.4.2.1.2 Handoff Decision

As well as keeping the signal quality, the VDL2 handoff will produce its overhead with triggering logic-link establishment and executing data handover. The overhead will cause the delay increase and capacity reduction in the process of handoff. To optimize the network performance, the decision of handoff moment is the key point: being too dull will worsen SNR and BER; being too sensitive will degrade system performance.

Taking the overhead into consideration, we applied a transition method in the handoff decision. The method is different from the absolute decision (in which the handoff will be triggered just after received signal strength is less than threshold) and supply a transition interval (SQP_PARA) as the handoff trigger threshold instead of the absolute threshold, as shown in Figure 5.13.

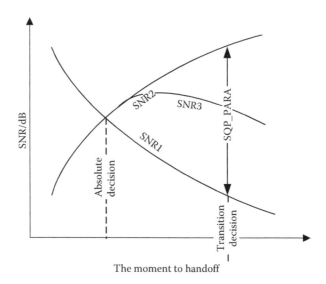

FIGURE 5.13
Threshold-based handoff decision.

The SINR-based handoff decision algorithm is as follows:

$$SQP_{newGS} \geq SQP_{originGS} + SQP\_PARA \qquad (5.25)$$

5.4.2.2 Experiment Setup

To simulate the handoff process, we allocated 3 VDL2 GS and 16 aircraft with fixed flight route in the simulation scenario, as shown in Figure 5.14.

In the scenario, we set the flight height to 10 km and the flight speed to 800 km/h. The GS antenna height is set to 30 m.

5.4.2.2.1 Simulation Configuration

The handoff simulations are configured as follows:

1. Packet birth in any node meets Poisson distribution for which Poisson intensity is 5.
2. Packet length meets uniform distribution from 0 bit to 8192 bits.
3. P-CSMA access probability: GS side is 90/256 and AC side is 13/256.
4. The maximum transmission attempt number M_1 is set to the default 135.
5. The DLS sublayer parameters are set by the defaults: the maximum number of retransmissions N_2 is set to 6; the length of sliding window is set to 4.

5.4.2.2.2 Performance Parameter

In practice, handoff will be triggered by one of the following conditions[19]:

Condition 1: The new GS signal quality is obviously larger than that of the origin GS.

Condition 2: N_2 timer is expired, where system will believe the connection invalid.

Condition 3: Since AC did not receive GS transmitted packet over a long time (default is 60 min), TG_2 timer is expired.

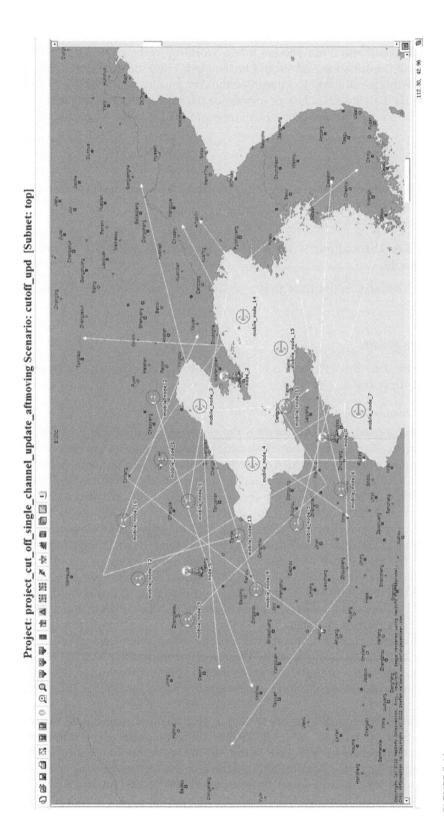

FIGURE 5.14
(See color insert.) Handoff scenario.

Among the three conditions, condition 1 is the active mode, working in the normal communication; conditions 2 and 3 are passive mode, working in the circumstance that cannot keep normal communication and should be avoided. Therefore, the first demand is to increase the active mode ratio and to guarantee the communication working in a normal state. We will define it as the first-class performance parameter. Besides, we will still consider the impact of handoff on overall performance and take it as the second-class performance parameter.

1. The first-class performance parameter

    Definition 5.5: Handoff ratio is the ratio of the passive handoff number to the general handoff number.

2. The second-class performance parameter

    Definition 5.6: Throughput is the ratio of successfully delivered bit rate and channel capacity.

    Definition 5.7: Delay is the time that a packet waits for before it is successfully delivered.

5.4.2.3 Simulation Results

The section will simulate the performance of the SINR-based transition handoff, verifying its feasibility and superiority.

5.4.2.3.1 Contrast Simulation

The purpose of this experiment is to verify if the proposed method can achieve obvious improvement compared to the SNR-based absolute decision method. We choose the first class as the performance parameter, which can reflect the handoff fundamental performance.

To make the contrast simulation, we modeled the two methods in VME module and then made the contrast simulations in the same scenario as Figure 5.14. We also added debugging statements in modeling so that we can observe the simulation objects such as target GS, handoff type, and signaling exchange from OPNET debugger (ODB) window.

At the beginning of contrast simulations, the link establishment will be triggered for all aircraft. Since the aircraft are well distributed at the initial time, we can assume that they have the same interference and the link establishment depends only on SNR, which is decided by the distance between AC and GS. The connection information (from ODB) shows that ACs of both decision methods are similar at the initial linking and tend to link to the GSs with the minimum distance.

With the developing simulation, the aircraft will fly along their respective tracks, and the node density will change continuously. We can observe that the handoff due to distance (between AC and GS) will undergo in both methods. The difference exists in that the SINR-based method will initiate handoff by excessive intensive density, but the SNR-based method will not.

The contrast curves of handoff ratio are shown in Figure 5.15. In the figure, SINR-based handoff curve is concave–convex with SQP_PARA ≈ 12 as the inflection point. ODB trace shows that with the increase in SQP_PARA, the SIR-induced handoff proportion will be continually falling in the active handoff and will be less than 50% when the curve arrives at the inflection point. At the same time, the number of SIR-induced handoffs is far higher than that of SNR-induced handoffs. Therefore, in the concave part, handoff is mainly

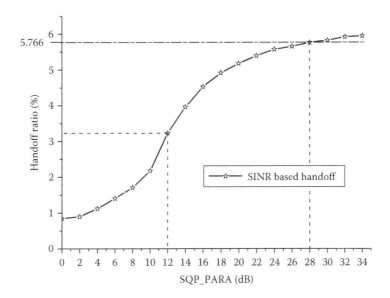

FIGURE 5.15
The SINR-based handoff versus signal strength-based handoff.

caused by the intensive density, and the frequency of active handoff is comparably higher; in the convex part, handoff is mainly caused by the distance (between AC and GS), and the frequency of active handoff is comparably lower. On the other side, the active handoff reduction will impair the system ability to allocate the radio resources and thus increase the passive handoff ratio.

On the whole, handoff ratio increases with the increasing of SQP_PARA. In a wide area (0–28 dB) that contains the inflection point, the SINR-based handoff rate is better than the SNR-based handoff rate for which the optimum value is 5.766 as shown in Figure 5.15. The improvement is derived from intensive density-based handoff trigger, which makes the handoff being adaptive. When SQP_PARA exceeds 28 dB, the SINR-based handoff ratio will continue to grow slowly, and handoff ratio tends to further deteriorate.

It should be pointed out that the handoff ratio curve will have a similar shape when the aircraft number or route is changed. But with the number increasing, SIR-based handoff will increase, and the inflection point will shift behind.

5.4.2.3.2 Transition Interval

This experiment will analyze the effect of SQP_PARA parameter on the handoff performance and optimize the threshold. We choose the second class as the performance parameter, which can reflect the handoff system performance. In a specified scenario, we deployed 24 group simulations taking SQP_PARA as the variable, and each group lasted 60 min. To eliminate the accidental factors of flight route, we made 50 scenarios with different flight route schemes. The average throughput and delay can be obtained from the multiple scenario-based simulations.

As shown in Figure 5.16, the average throughput curve has a quasi-parabolic shape with downward opening. The average throughput will first increase, then decrease with SQP_PARA growing up. It will reach the maximum value when SQP_PARA is about 8 dB. On the contrary (as shown in Figure 5.17), the average delay curve has a quasi-parabolic shape

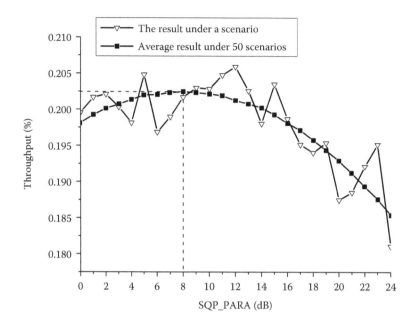

FIGURE 5.16
Throughput versus SINR threshold.

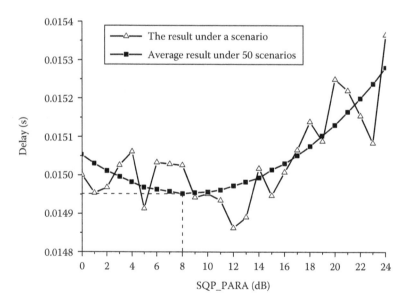

FIGURE 5.17
Delay versus SINR threshold.

with upward opening. The average delay will first decrease, then increase and will also reach the minimum value at SQP_PARA ≈ 8 dB.

In this experiment, throughput and delay reflect the good consistency, namely, they can simultaneously reach the optimal values at a point. The point is the optimal handoff threshold, denoted as SQP_PARA$_{opt}$. At the same time, the experiment shows that SQP_PARA$_{opt}$

always falls in the area in front of the inflection point. The result indicates that SIR-based handoff is dominant at the optimal point.

5.5 Conclusion

The chapter first discusses the application prospect of the VDL2 system, stating the technical advantage of the VDL over the current ACARS and the necessity to realize the system transition. Then, the four VDL modes are outlined. By comparison, the conclusion is drawn that the VDL2 will dominate the air–ground VHF data link for the ATN.

Focusing on the VDL2, the three-layered architecture is introduced on the whole, and it is developed in detail by physical layer, data link layer, and subnetwork layer. We still analyzed the key technologies in the VDL2 realization, involving D8PSK modulation, RS channel encoding, P-CSMA, AVLC, handoff, etc. On the basis of the protocol and technology, an entire VDL2 simulation model is built on the OPNET platform.

Last but not least, a handoff decision method is put forward based on SINR from the view of co-interference and handoff overhead. To testify its superiority and feasibility, a VDL2 handoff model is established on OPNET taking dynamic signaling interaction, handoff timing, mobile node, multifrequency access, etc., as the modeling mechanisms. Simulation experiments prove that the proposed method can realize handoff according to the flight nodes density and reduce the passive handoff ratio effectively. Further simulations also show that there exists optimal handoff threshold that can make the performance such as throughput and delay to be optimized, which supplies a way to improve the current handoff method.

References

1. Hui-yuan Z, Xue-jun Z, Jun Z. Migration from ACARS to new VHF data link in China. *Proceedings of International Conference on ITS Telecommunications*, Chengdu, China, 2006: pp. 1297–1300.
2. Kitaori J. A performance comparison between VDL mode 2 and VHF ACARS by protocol simulator. *Proceedings of 28th Digital Avionics Systems Conference*, Orlando, FL, 2009: pp. 4.B.3-1–4.B.3-8.
3. ICAO. Annex 10 aeronautical telecommunications. Digital data communication systems. Montreal, Quebec, Canada: ICAO, 2007.
4. James W R. Are we nearer to NEXCOM? *Avionics Today* (December 1, 2003). http://www.aviationtoday.com/av/commercial/Are-We-Nearer-to-NEXCOM_1201.html#.VQku_tKl_YU. Accessed: August 13, 2014.
5. RTCA. Signal-in-space minimum aviation system performance standards (MASPS) for advanced VHF digital data communications including compatibility with digital voice techniques, RTCA Standard DO-224A. Washington, DC: RTCA, 2000.
6. Bretmersky SC, Murawski RW, Konangi VK. Characteristics and capacity of VDL mode 2, 3, and 4 subnetworks. *Journal of Aerospace Computing, Information and Communication*, 2005, 2(11): 470–489.

7. ICAO. VHF Digital Link (VDL) (Mode 2) Standards and recommended practices, AMCP/4-WP/70. Montreal, Quebec, Canada: ICAO, 1996.

8. ICAO, Annex 10. VDL-2 Standards and recommended practices. Montreal, Quebec, Canada: ICAO, 1997.

9. Sklar B. *Digital Communications-Fundamentals and Applications*. Beijing, China: Electronic Industry Press, 2002.

10. Xuehong C, Zongcheng Z. *Information Theory and Coding*. Beijing, China: Tsinghua University Press, 2004: pp. 145–147.

11. Lin G, Zhi-jun W. A new method to analyze asynchronous P-CSMA. *Journal of Computers*, 2014, 9(1): 161–167.

12. ISO/IEC. Information technology—Telecommunications and information exchange between systems—High-level data link control procedures, IEC 13239. ISO, 2002.

13. Whyman AW, Adnams MG. Proposed guidance material in support of VDL handoffs, AMCP WG M. Malmo, Sweden: AMCP, 2000.

14. Lin, G, Zhi-jun, W. Performance analysis of MAC in VDL2 based on OPNET simulation. *Applied Scientific Research and Engineering Development for Industry*. Zurich-Durnten, Switzerland: Trans Tech Publications, 2013: pp. 1643–1646.

15. Li X, Ye M. *Network Modeling and Simulation with OPNET Modeler*, Xian Electronic Science and Technology University Press: Xian, China, 2006: pp. 164–169, 212–214.

16. Jiasheng W, Jiahun, L. *Stochastic Process*. Tianjin, China: Tianjin University Press, 2003: pp. 26–31.

17. Bretmersky S C, Konangi V K, Kerczewski R J. Performance of VDL mode 2 for the Aeronautical Telecommunications Network. *Proceedings of 21st Digital Avionics Systems Conference*, Los Alamitos, CA, 2002: pp. 3.C.5.1–3.C.5.9.

18. Lin G. Key technology research of VDL2 aeronautical data link. Tianjin, China: Tianjin University, 2014: pp. 90–96.

19. Desperier B, Esposito M, Delhaise P. VDL mode 2 hand-off analysis and simulation, AEEC DLK Systems Subcommittee. Tampa, FL: AEEC, 2004.

6

GNSS Multipath Interference and Mitigation for UAVs in Urban Canyon Environments

Saeed M. Khan

CONTENTS

6.1 Introduction

Unmanned aerial vehicles (UAVs) flying through urban canyons are subject to multipath signals as a result of reflection, refraction, and diffraction of satellite signals traveling to the receiver; this can lead to positional inaccuracies in this time-critical operation that can lead to crashes. Indeed, analytical models show that for a GPS L1-C/A (coarse or clear signal/acquisition), the time to detect a satellite signal increases considerably under conditions of multipath interference (Schmid and Neubauer 2004, 503–509; Heng et al. 2014; Xie and Petovello 2015). Traditionally, two different approaches have been taken to dealing with the multipath problem. One approach involves signal processing and the other involves designing an antenna (spatial processing) capable of rejecting multipath signals (Rahmat-Samii 2003, 265–275; Nedic 2009; Kos et al. 2010, 399–402). While several signal processing solutions for dealing with multipath impact exist, they require advanced receiver architectures that either mitigate the effect of multipath signals or use signals identified as multipath to enhance location accuracy. A question that remains to be answered is whether or not acquisition time for satellite signals can be optimized by carefully matching antenna and receiver.

The purpose here is to suggest an antenna–receiver combination that is designed to cut out multipath signals and to rely on line of sight (LOS) signals for location information. Even with less than four LOS satellite signals in any one global navigation satellite system (GNSS), the mobile unit can be located by using available receiver architectures that support multiple satellite navigation systems, for example, combining GPS and Galileo LOS signals to determine position. The proposed solution employs an antenna design that rejects multipath by cutting off signals that arrive at elevation angles low to the horizon; it works with different receiver architectures and does not require additional processing time for the purpose of multipath mitigation. Preliminary analysis has shown that the conical ring antenna can outperform most antenna solutions in dealing with the multipath problem. Some common solutions rely on the use of choke rings (Basilio et al. 2005, 233–236), frequency selective surfaces (FSS) or photonic band gap (PGB) surfaces (Rahmat-Samii 2003, 265–275), resistive ground planes, and the reduced surface wave (RSW) antennas (Basilio et al. 2005, 233–236). In addition, the ring structure expected to provide protection from jamming and spoofing signals incoming at angles low to the horizon.

It is expected that this treatment will provide the readers with a general physical understanding of the multipath phenomena and its importance to unmanned systems. The chapter begins with a general introduction to the multipath environment that impacts both navigation and communication of unmanned aircraft systems (UAS) in an urban canyon setting. The work will target both practitioners and researchers interested in reviewing the fundamentals of interference and important considerations for multipath mitigation. Following the review, the special case of GNSS multipath interference in urban canyon environments will be provided starting with background information about the evolution of GPS and its interoperability with the Galileo system under the conditions of continued modernization. Next, the author will present the results of his own research to a spatial mitigation technique that capitalizes on the multiple GNSS systems to cut down on the negative impact of multipath. It is expected that this treatment will provide the readers with a general physical understanding of the multipath phenomena and its importance to unmanned systems.

6.2 Multipath Phenomenon

Although radio frequency interference, multipath, and ionospheric scintillation are all potential sources of error, technological advancements in GNSS systems have diminished the impact of most sources of error to the point where multipath and shadowing (a phenomenon where the direct path suffers excess attenuation) remain as significant issues (Ward et al. 2006, 279–295). The multipath phenomenon is experienced by a receiver when it receives either a direct path signal from a source along with one or more reflected/diffracted satellite signals. If one visualizes these signals to be following the path of the rays in Figure 6.1, then the receiver is being subjected to multipath signals that can lead to errors. In this simplified figure, multipath signals arrive at the receiver through reflected

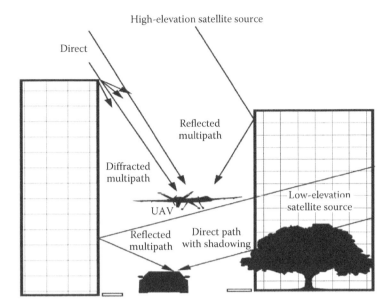

FIGURE 6.1
Direct/multipath headed to a receiver on a car and a UAV. Only one satellite signal is used in each case.

and diffracted angles to the UAV from a single high-elevation satellite source. A low-elevation satellite source signal is received at the car by both a multipath reflected signal and a direct signal that has undergone shadowing while passing through foliage. The strength of the shadowed direct signal is attenuated and in some cases may even be lower in level than the multipath signal and this can cause problems in discrimination between the signals at the receiver. It should be noted that we have ignored the compound effect of all satellites in view that should be affecting both receivers.

In order to understand the impact of multipath on receivers, it is important to review how distances to satellites are measured. Distances to the satellite vehicle (SV) from receiver can be measured by cross-correlating the transmitted pseudorandom (PRN) code to a known PRN for the particular SV at the receiver until the phases of both codes match. If the clocks on the SV and the receiver can be synchronized, then the time shift Δt required to match phases can be multiplied with the speed of light to gives us the distance ($c \cdot \Delta t$). Errors are introduced in the calculation of this time shift in a multipath environment, and this can lead to positional inaccuracies.

Not all multipath signals introduce errors. Multipath signals travel different path lengths to get to the receiver and can be broken down into near echoes and far echoes. If the time of arrival between two echoes is greater than the inverse of the system bandwidth, then they can be resolved (Jahn et al. 1996, 1221–1226). Therefore, multipath delays are greater than twice the spreading code period they are resolvable (Ward et al. 2006, 279–295), and as such multipath from nearby objects (near echoes) can distort the correlation functions. For example, if a receiver uses C/A-code at a chipping rate of 1.023 MHz multipath with path lengths greater than 600 m longer, then the direct signal path can be resolved. Problems can also stem from grazing reflected signals arriving from greater distances as they tend to destructively interfere with the direct signal by being in phase opposition. The impacts of different ray paths are summarized in Table 6.1.

TABLE 6.1

Ray Path and Impact

Ray Path	Distance/Travel Time	Path Characteristics	Impact Comment
Direct path to receiver w/o shadowing	Direct with no attenuation in channel	Unobstructed	Does not introduce multipath errors.
Direct path to receiver with shadowing	Direct with attenuation in channel	Obstructed by objects that cause attenuation in signal	Excessive attenuation can cause signal strength fall below multipath signals causing resolving issues.
Multipath near echoes	Arrive within twice the spreading code period after direct signal	Reflected/diffracted one or more times	Distorts the correlation function received composite signal and locally generated reference.
Multipath far echoes	Arrives taking more than twice the time for the spreading code period following direct signal	Reflected/diffracted one or more times	Resolvable from near echoes in receiver.
Grazing reflected rays At low angles to the horizon	Almost the same time taken as the direct path	Reflected once	In phase opposition to direct path; can potentially cancel out the direct path ray.

Source: Ward, P. et al., Interference, multipath, and scintillation, in *Understanding GPS Principles and Applications*, 2nd edn., E. Kaplan and C. Hegarty (eds.), Artech House, Norwood, MA, 2006, pp. 279–295.

6.3 Evolution of GPS and Interoperability with the Galileo System

Since the 2004 GPS–Galileo agreement establishing cooperation between the U.S. and EU member states, eight major documents describing the path and course of this cooperation have come out. Table 6.2 is a summary of these documents and describes the historical progression of this collaboration. It is fairly easy to see that the path forward includes, among other things, optimized civil signals that would allow future receivers to track GPS and/or Galileo systems with higher accuracy. The U.S. government has also started cooperation conversations with India, China (most recent), Japan, Australia, and Russia (currently on hold).

6.3.1 Modernization and New Civil Signals for GPS

Another bit of good news for civilian users is that the U.S. government has started to bring into the picture three new signals for their use, which includes the L2C, L5, and L1C. The L1C capabilities will start to come online in 2016 with GPS III phase and be available on 24 satellites around 2026 (U.S. Air Force 2014b). Modernization plans for both GPS and Galileo will make it possible for upgraded receivers to track L1C GPS and/or E1 Open Service Galileo signals with higher accuracy. The following table summarizes the modernization status of these new civil signals in Table 6.3.

6.3.2 Impact of Evolution and Modernization on Urban Canyon–Type Environments

The previous discussion of modernization and evolution of GNSS systems presents some new opportunities for the spatial processing of multipath signals in an urban canyon. Following is an itemized list of some of these points:

TABLE 6.2

Historical Progression of the GPS–Galileo Cooperation

Year and Document	Key Points
2004 GPS-Galileo Agreement (U.S. Air Force 2004)	Framework for cooperation between the Parties in the promotion, provision and use of civil GPS and GALILEO navigation and timing signals and services, value-added services, augmentations, and global navigation and timing goods. The Parties intend to work together, both bilaterally and in multilateral fora, as provided herein, to promote and facilitate the use of these signal, services, and equipment for peaceful civil, commercial, and scientific uses, consistent with and in furtherance of mutual security interests.
2006 Joint Statement on Galileo-GPS Signal Optimization (U.S. Air Force 2006)	EC and the United States signed the Agreement on the Promotion, Provision and use of Galileo and GPS Satellite-Based Navigation Systems and Related Applications on the compatibility and interoperability of the Galileo and GPS satellite navigation systems. A central element of the Agreement was a common baseline signal structure that could be optimized for greater performance.
2007 Joint Statement on GPS-Galileo Working Group B (U.S. Air Force 2007)	Agreement on the Promotion, Provision and Use of Galileo and GPS Satellite-Based Navigation Systems and Related Applications. The purpose of Working Group B is to consider, inter alia, non-discrimination and other trade related issues concerning civil satellite-based navigation and timing signals or services, augmentations, value-added services, and global navigation and timing goods.
2007 Joint Fact Sheet on GPS-Galileo Cooperation (U.S. Air Force 2007)	The United States and the European Union are designing GPS and Galileo to transmit one or more common civil signals, so future users will enjoy the benefits of multiple PNT satellite constellations. Benefits include increased satellite availability (particularly in urban environments) and improved resistance to signal interference. Signals will be provided on two common frequencies by the United States and the European Union without direct user fees.
2007 Joint Press Release on Civil Signal Design (U.S. Air Force 2007)	The resulting GPS L1C signal and Galileo L1F signal have been optimized to use a multiplexed binary offset carrier (MBOC) waveform. Future receivers using the MBOC signal should be able to track the GPS and/or Galileo signals with higher accuracy in challenging environments that include multipath, noise, and interference.
2008 GPS-Galileo Working Group B Meeting (U.S. Air Force 2008)	The Working Group addressed specific questions about Galileo raised by U.S. industry and considered proposals for enabling global equipment manufacturers to develop and test GPS/Galileo receivers and applications on a level playing field that benefits users.
2008 Joint Statement first plenary meeting (U.S. Air Force 2008)	The improved common civil signal, referred to as L1C on GPS and E1 Open Service on Galileo, has been optimized using a Multiplexed Binary Offset Carrier (MBOC) waveform. Future receivers using this signal should be able to track the GPS and/or Galileo signals with higher accuracy in challenging environments.
2010 Joint Statement on Combined Performance of GPS and Galileo (U.S. Air Force 2010)	The combination of GPS and Galileo services provided noteworthy performance improvements particularly in partially obscured environments, where buildings, trees or terrain block large portions of the sky. Dual-frequency receivers provide additional improvements in most environments.
2013 Interim Report of ARAIM Technical Subgroup (U.S. Air Force 2013)	Working Group C (WG-C) is designed to enhance cooperation for next generation GNSS. One of the objectives of WG-C is to develop integrated applications for Safety-of-Life services. To this end, WG-C established the ARAIM Technical Subgroup. The objective of the Subgroup is to investigate ARAIM (Advanced Receiver Autonomous Integrity Monitoring). The further goal is to determine whether ARAIM can be the basis for a multi-constellation concept to support air navigation worldwide.

TABLE 6.3

New Civil Signals

Signal	Frequency and Modulation	Launching Dates	Key Feature
L2C (second GPS civilian signal)	1227.6 MHz; biphase shift key (BPSK) operation	Started launching 2005 and has 14 GPS satellites as of October 7, 2014; 24 GPS satellites available around 2018	First to provide dual frequency capabilities to civil signals, leading to higher-precision locational capabilities than with the L1 C/A alone
L5 (third GPS civilian signal)	1176.45 MHz; biphase shift key (BPSK) operation	Started launching 2010 and has 7 GPS satellites as of October 7, 2014; 24 GPS satellites available around 2021	Greater bandwidth for improved jam resistance; higher transmitted power than L1 C/A and L2C
L1C (fourth GPS civilian signal)	1575.42 MHz; Multiplexed Binary Offset Carrier Modulation	Begins launching with GPS III in 2016 moving to 24 satellites with capability in 2026	Designed for international GNSS interoperability

Source: U.S. Air Force, New civil signals, 2014a, http://www.gps.gov/systems/gps/modernization/civilsignals/#L1C, accessed November 9, 2014.

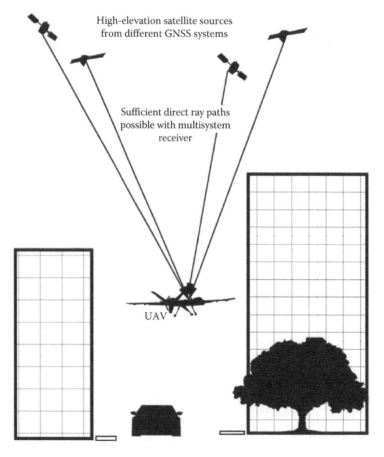

FIGURE 6.2
Multisystem receiver would benefit from more overhead SVs in an urban canyon.

1. The development and maturing of more and more GNSS (Table 6.2) systems over time will increase the focus on developing multisystem receivers allowing receivers to take benefit of more unobstructed direct line signals in an urban canyon-like environment (Figure 6.2).

2. The new developments bring up prospects for a spatial processing system that cut of all signals from SVs that are not at high elevation.

3. The coming on line of new civilian signals L2C, L5, and L1C will allow higher levels of precision (Table 6.3) even if low-elevation satellites are cut out, since additional signals would be available and at different frequencies.

4. The cutting off the low-elevation SVs will cut off grazing reflected rays and near echoes at these angles as well (see Table 6.1).

5. Based on the trajectories and cooperation between countries in the area of GNSS, there will be increased focus on the design and development on multisystem receivers.

6. Based on items 1–5, one can see that the demand for spatial processing antennas for use in urban canyon-like environments will increase and also will be supported by an availability of new multisystem receivers.

6.4 Establishing the Characteristics of an Antenna for Urban Canyon

GNSS antenna design considerations that respond to the need for urban canyon-like environments can be established by using the evolution and modernization trajectories alongside important electromagnetic considerations of good bandwidth (especially important when using multiple systems), axial ratio, and radiation pattern. To begin our discussion, we focus on some different multipath-rejecting antennas that are quite well known and the desirable characteristics that they exhibit but are not necessarily ideal for UAV application.

6.4.1 Choke Ring Antenna

While the choke ring antennas have been primarily used for precision in measuring geodetic references, it is has established itself as a well-known survey antenna and has many known variations (Basilio et al. 2005, 233–236; Sciré-Scappuzzo and Makarov 2009, 33–45). The basic structure of the antenna consists of a number of concentric metallic cylinders that sit on top of a ground plane, creating a corrugated surface (Figure 6.3 illustrates the basic structure of a choke ring antenna). The surface prevents surface waves and plane waves from creating edge radiation and thereby reduces backlobe radiation and radiation at low elevation angles from the horizon. This antenna also has good phase center (the apparent point from which the electromagnetic wave spreads spherically; it should be noted that this is not a single point but varies in position based on the angle of incidence) characteristics, that is, small variation with the angle of incidence. The depth of the corrugation should be greater than quarter wavelength of its lower operational frequency and less than half the wavelength of its upper operating frequency (about 2.5 in. for an L1/L2 antenna). The size of the ground plane is also an important consideration (about 15 in. for an L1/L2 antenna). The size and weight of this antenna make it an unlikely choice for an UAV.

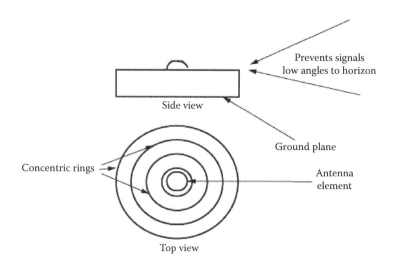

FIGURE 6.3
Choke ring antenna.

6.4.2 Frequency Selective Surfaces (FSS) or Photonic Band Gap (PGB) Surfaces

FSS or PGB surfaces when used as ground planes have been used to cut down on surface waves and edge radiation by offering high impedances to surface currents on them. This high impedance property works only within a design bandwidth (Sievenpiper et al. 1999, 2059–2073).

The goal of FSS is to come up with smaller, thinner, and lighter ground planes for antennas that broaden the possibility of deployment in many more applications when compared with survey antennas like the choke ring antenna. Generally, these FSS are constructed by placing of elements on top of substrates spread out like an array. While many different element structures exist, such as variations of square loops and split rings, these elements are carefully designed such that their interaction with the ground plane and with each other produces the proper capacitive gaps and inductive traces to create resonances (high impedance) to surface currents. The elements can be as small as $\lambda/12$ (Bayatpur and Sarabandi 2007, 1239–1245; Bayatpur 2009, 1–164) and are not considered to be sensitive to the angle of incidence. The elements can be placed very close to the antenna and this can help reduce the dimensions of an optimal design (Figure 6.4 shows an FSS surface with antenna).

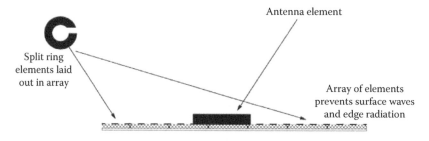

FIGURE 6.4
FSS and antenna.

6.4.3 Resistive Ground Planes

Another method of shaping the radiation pattern to avoid multipath is to use resistive cards at the edges of the antenna ground plane (Rojas et al. 1995, 1224–1227). The resistive edges (Figure 6.5) of the ground plane prevent diffraction and help shape the radiation pattern such that it offers less gain to signals arriving at low angles to the horizon. The resistive edges, helps prevent distortion of the phase center due to this edge effect.

6.4.4 Reduced Surface Wave Antennas

RSW antennas have a promising design for rejecting multipath (Basilio et al. 2005, 233–236). The shorted-annular-ring-reduced surface wave antenna consists of a circular patch that shorted to the ground plane at a carefully selected distance from the center that is based on the analysis required to prevent lateral radiation. The performance of this antenna when sitting on a 35.6 cm ground plane proved to be promising when compared to the choke ring antenna pattern gain on the horizon (10 dB lower and better at cutting off multipath arriving at angles low to the horizon). This antenna would prove to be at least an order of magnitude lighter than the choke ring.

6.4.5 Conical Ring Antenna

The conical ring antenna (Khan 2014, 232) has been designed especially for UAVs in urban canyon following the design criteria listed as follows:

1. Design should compare favorably or exceed in multipath rejecting capabilities of choke ring antennas; antennas using FSS or PGB surfaces; antennas using resistive edges on their ground planes; and RSW antennas.

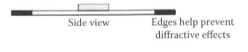

Side view   Edges help prevent
                   diffractive effects

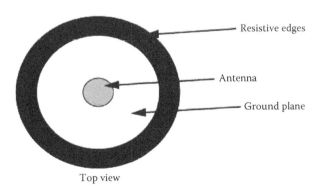

FIGURE 6.5
Antennas with resistive edges on their ground planes.

2. Should be able operate at or be extendable to both L1/L2 frequencies. Such that it can take advantage of the L1C/L2C and the E1 Open service for Galileo.

3. It should have the size, weight, and shape that make it a deployable payload for both small and large UAVs.

The structure of the conical ring antenna is shown in the Figure 6.6. The lowest layer of the antenna is a feed structure for this multifeed antenna. The feed structure is responsible for creating four equal amplitude excitations that are phased 0°, 90°, 180°, and 270° to produce right circularly polarized radiation (RHCP). Four probes carrying these excitations emerge from below a metallic ground plane and pass through ground plane below the patches without touching either one. The feeding posts also serve to mechanically support the ring and patch combination. Once it passes through the substrate of the patch, it makes contact with the top patch. In the event a dual frequency (say L1/L2) design is used, the probes would pass through the bottom patch without making contact with the top patch. The conical ring is supported by a dielectric structure built by a 3-D printer.

The model conical ring is backed by a dielectric material called abs plastic (relative dielectric constant 2.87). Other dielectric materials (polyethylene, fiber glass, ceramics, etc.) can also be used, provided the dimensions of other components are scaled properly.

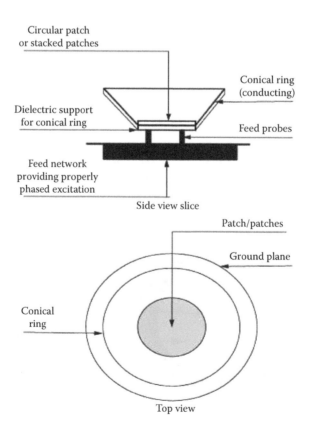

FIGURE 6.6
Construction of conical ring antenna (not to scale).

TABLE 6.4

Gain Comparison with Different Solutions at Elevation Angles Close to the Horizon

Angle in Degree	90	85	80	75	70
RSW (measured (Basilio et al. 2005, 233–236))	−30 dB	−25 dB	−21 dB	−15 dB	−11 dB
Conical ring (simulated)	−31 dB	−29.5 dB	−25 dB	−21 dB	−17.5 dB
Choke Ring (measured (Basilio et al. 2005, 233–236))	−18 dB	−16 dB	−15 dB	−13 dB	−10 dB

A smaller value of gain, at these elevation angles, indicates better multipath rejection.

Currently, the ground plane is about 16 cm in diameter, post height is about 1 cm above ground plane, circular patch has a diameter of about 7.2 cm. Increasing the dimensions of either the conical ring and/or circular patch inside will lower the resonant frequency. Using a material of higher dielectric constant will also lower resonant frequency.

When compared with choke rings (Basilio et al. 2005, 233–236), the conical antenna demonstrates superior multipath rejection in all cases of elevation angles between 70° and 90° (or angles low to the horizon) as evidenced by the pattern gain. The conical ring gains were between 8 and 13 dB lower for the field pattern. The largest dimension of conical antenna without a radome is about 16 cm whereas the largest dimension for choke rings is around 30 cm. There will also be a considerable difference in weight which makes the conical ring also more desirable from a payload point of view.

When the RSW antenna (Basilio et al. 2005, 233–236) sits on a 35.6 cm ground plane, the conical antenna demonstrates superior multipath rejection in all cases of elevation angles between 70° and 90° (or angles low to the horizon) as evidenced by the pattern gain. Regardless of the RSW antenna itself, for these results it was sitting on a large ground plane more than four times the largest dimension of the conical ring. While the choke ring might be an order or more heavier than the RSW plus ground plane combination, the conical ring beats them both in the desirable size and weight criteria.

While antennas with resistive ground planes have not been compared against the conical ring, the paper describing the E-plane pattern (Rojas et al. 1995, 1224–1227) in 70°–90° elevation finds that resistive ground provides a minimum gain that is greater than −13 dB. The best and worst case performance for the conical ring in this range is −31 dB (at 90°) and −17.5 (at 70°) indicating superior performance in this case as well (Table 6.4).

6.5 Comparisons of Four Related Designs

In order to study how structural variations in conical ring antenna impact the antenna and in order to conduct further analysis of its behavior, four different antenna structures were compared. These structures include a circular patch, circular patch with FSS surface, the conical ring antenna (which is simply a circular patch surrounded by a conical ring), and a conical ring surrounded by an FSS surface. The dimension of the ground plane is kept the same in all cases and each structure is fed using four feed signals having equal amplitudes and phase to deliver.

To start, a simulation for the radiation pattern for all four are plotted for comparison (Figure 6.7) and their multipath rejecting capabilities are compared (Table 6.5). From Table 6.5, comparing a plain circular patch with the circular patch that has been surrounded by FSS elements, it can be seen that the latter provides a better multipath rejecting

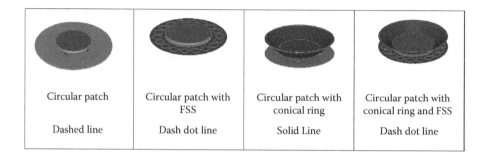

Circular patch	Circular patch with FSS	Circular patch with conical ring	Circular patch with conical ring and FSS
Dashed line	Dash dot line	Solid Line	Dash dot line

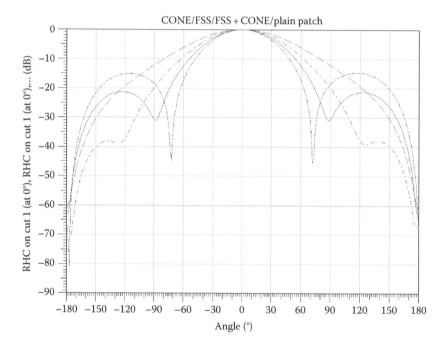

FIGURE 6.7
(See color insert.) Patterns of all four structures compared.

TABLE 6.5

Gain Comparison with Different Solutions at Elevation Angles Close to the Horizon—RHC Pattern

Angle in Degree	90	85	80	75	70
Circular patch	−12 dB	−11 dB	−10 dB	−9 dB	−8 dB
Circular patch with FSS surface	−20 dB	−18 dB	−16 dB	−14 dB	−12 dB
Conical ring	−31 dB	−29.5 dB	−25 dB	−21 dB	−17.5 dB
Conical ring with FSS	−18.5 dB	−21 dB	−24 dB	−32.5 dB	−33 dB

A smaller value of gain, at these elevation angles, indicates better multipath rejection.

capability at close to horizon angles. This is evident from the table as the circular patch with FSS elements provides 8 dB lower gain at 90° and 4 dB lower gains at 70°.

When a comparison is made between the conical ring and the circular patch with an FSS surface, the conical ring shows superior multipath rejection capabilities by having anywhere between 6 and 11 dB less gain. An interesting situation arises when the conical ring is compared with the conical ring plus FSS. While the conical ring does outperform the conical ring with FSS in the 70°–90° range, there is a range around 60° when the one with FSS has better rejection. A question that this raises is whether or not it is possible to tune these behavior-changing elements in the FSS array.

6.5.1 Impact of the Finite Ground Plane in the Absence of a Conical Ring

We know that for an ordinary circular patch, the radiation pattern, directive gain, and input impedance vary widely with the size of the ground plane. Analytical studies (Bhattacharyya 1990, 152–159) have shown that for a circular patch with a ground plane radius of $0.63\lambda_0$ (where λ_0 is the free space wavelength), the patch antenna achieves maximum gain. Large ground planes are undesirable for most applications, especially when designed for small UAVs.

6.5.2 Effectiveness of the Conical Ring

With the inclusion of the conical ring (Khan 2014, 232), the RMS electrical field distribution, RMS magnetic field distribution, and the time average power density in the near field (Figures 6.8 through 6.10), when compared to the plain patch, appear much more symmetrical and tightly bounded to the phase center. (In practical terms, when referenced to the phase center, the fields radiated by the antenna are spherical waves with equiphase surfaces over most of the angular space in the main beam.) The phase center for the conical ring antenna is located 1 mm above the ground plane as per simulation results; the field and power distribution is from a slice taken at 7 mm above the ground plane.

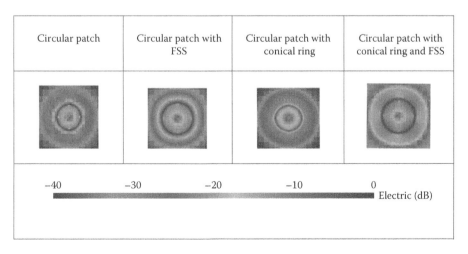

Circular patch	Circular patch with FSS	Circular patch with conical ring	Circular patch with conical ring and FSS

−40   −30   −20   −10   0
                                                    Electric (dB)

FIGURE 6.8
(See color insert.) RMS electric field distributions in a slice taken at 7 mm on top of the ground plane at 1.575 GHz.

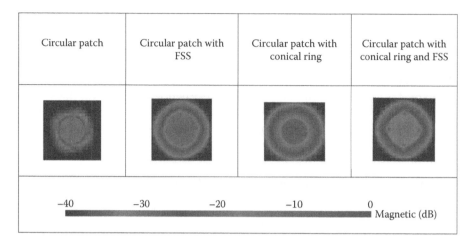

FIGURE 6.9
(See color insert.) RMS magnetic field distributions in a slice taken at 7 mm on top of the ground plane at 1.575 GHz.

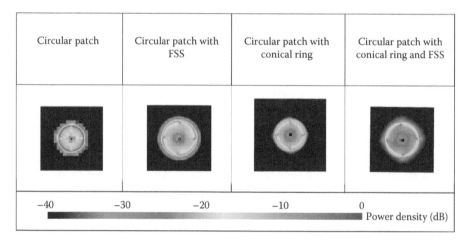

FIGURE 6.10
(See color insert.) Time average power density in a slice taken at 7 mm on top of the ground plane at 1.575 GHz.

In the previous sections multipath rejection benefits of the conical ring antenna were noted. The tightly held near field should require a smaller ground plane making it possible for the antenna dimensions to be further reduced. Analytical, numerical, and experimental studies are needed to investigate the impact of the ring on size of ground plane.

6.5.3 Impact of FSS Elements

Observations of field and power distributions for all four antennas from simulation (Figures 6.7 through 6.10) indicate that the use of FSS surfaces in combination with the conical ring has an interesting radiation pattern that allows a null to form further away from the horizon (at a smaller elevation angle $\Theta = \pm70°$ horizon is at $\Theta = \pm90°$ elevation). Figures 6.8 through 6.10 show how the circular patch with FSS elements surrounding it creates a more symmetrical field and power distribution (with or without the conical structure) when compared with an ordinary circular patch. With a conical structure added on the RMS electric field distribution seems to spread further toward the edges of the ground plane.

Characteristics of FSS surfaces are almost entirely determined by FSS elements, which in this case is the ring structure. One of the properties of resonant length standard FSS elements is its reflection and refraction depend on the angle of incidence (*Note*: This is not the case for smaller nonresonant lengths). Therefore, by changing the FSS element structure it should be possible to change the null location on their radiation patterns. This impact needs to be further investigated as it provides an additional degree of freedom to control where nulls form.

6.5.4 3-D RHC and LHC Pattern of Conical Rings

Figures 6.11 and 6.12 shows the 3-D RHC and LHC patterns, respectively. The RHC pattern shows a maximum gain of about 9.3 dBi (decibels isotropic). The pattern also shows very low gain for elevations that corresponds to low angles to the horizon and primary sources. The symmetrical nature of the patterns is both a function of the symmetry of the antenna and the progressively phased equal amplitude feeds. The 3-D LHC pattern also shows very low gain for near horizon and this is a plus as well in the event that there is a change in the handedness of the reflected signals.

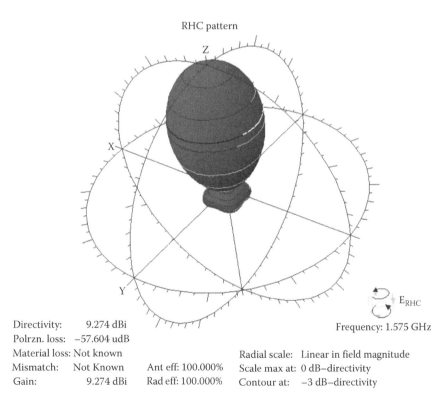

Directivity:	9.274 dBi		Frequency: 1.575 GHz
Polrzn. loss:	−57.604 udB		
Material loss: Not known			
Mismatch:	Not Known	Ant eff: 100.000%	Radial scale: Linear in field magnitude
Gain:	9.274 dBi	Rad eff: 100.000%	Scale max at: 0 dB–directivity
			Contour at: −3 dB–directivity

FIGURE 6.11
(See color insert.) 3-D RHC radiation pattern.

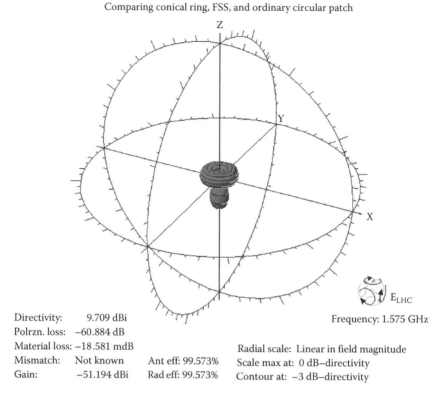

Comparing conical ring, FSS, and ordinary circular patch

Directivity: 9.709 dBi
Polrzn. loss: −60.884 dB
Material loss: −18.581 mdB
Mismatch: Not known   Ant eff: 99.573%
Gain: −51.194 dBi   Rad eff: 99.573%

Frequency: 1.575 GHz

Radial scale: Linear in field magnitude
Scale max at: 0 dB−directivity
Contour at: −3 dB−directivity

E_{LHC}

FIGURE 6.12
(See color insert.) 3-D LHC radiation pattern.

6.6 Future Work

Some of the unresolved issues of the conical antenna are as follows:

1. While the dimension of the conical antenna is much smaller than other well-known multipath rejecting types, and while it compares favorably with or exceeds the capabilities of these antennas, its largest physical dimension at 16 cm is still quite large for some smaller UAVs. Future work should explore the impact of using a higher dielectric constant substrate and support materials and the ability to scale down the size of the path, conical ring.

2. Although the system is designed to accommodate stacked patches allowing for dual frequency operation (e.g., L1/L2), this has not been studied yet. Stacking to patches on top of one another where the feed probe comes in without connecting with the patch and then makes contact with the top patch is used in many applications (Figure 6.13 illustrates this concept). The two patches are parasitically coupled and provide two distinct resonances and can be tuned to the dual frequencies being covered.

3. In Table 6.1, it was mentioned that not all multipath rays cause positional errors. Rays that arrive after the direct ray by more than two chip code periods does

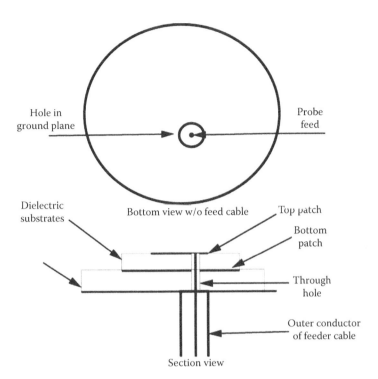

FIGURE 6.13
Stacked patches.

not create multipath errors since they can be resolved from multipath within this limit. So nearby reflection is more important in this case. These reflections might be coming from the aircraft itself or other aircraft flying in close formation. The proper placement of the antenna on the aircraft body might solve this problem, since we are now talking about signals coming from high angles as the low angles are afforded protection by the ring structure. A reflected ray is likely to cause a problem if it reflects onto the antenna from a higher point than the antenna. Keeping the antenna properly spaced from those at higher points might avoid multipath (Figure 6.14) from the aircraft body.

When it comes to close formation flights, the problem does not go away as easily as there is no guarantee that the reflections are not coming at high angles from other aircraft (Figure 6.15). There are good reasons and opportunities to explore further what happens in close formation and how to arrive at an ideal placement of antennas on an aircraft.

4. Finally, for future work should test antenna in the multipath environments resembling an urban canyon. As it is not possible to fly UAVs without a certificate of authorization (COA) from the FAA, a ground vehicle can be used for testing. Driving a truck through a metropolitan downtown with the test antenna hooked onto a multisystem receiver (e.g., SX-NSR) while conducting GPS only, Galileo only, and dual-tracking tests is critical, with and without the use of a hybrid system. Figure 6.16 shows how these systems will be mounted on trucks. Note that the antenna is actually mounted on an UAV and the combination is strapped

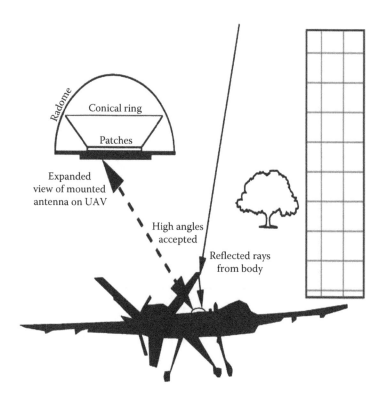

FIGURE 6.14
Antenna receiving high angle multipath signals from aircraft body.

onto a truck. In the event a COA is available, one can proceed with such a test in an authorized area with the antenna and receiver as payload (Figure 6.17 shows an UAV that might be able to carry such a payload).

6.7 Summary and Conclusions

This chapter discusses the key issues associated with GNSS positioning accuracy in an urban canyon–type environment where multipath can lead to errors and errors can lead to catastrophic situations for UAVs and their payloads. Spatial processing solutions to multipath are being preferred here to ones that use signal processing. More importantly, one particular spatial processing antenna is being forwarded, both at satisfying specifications and at establishing a design criteria that was used to come up with these specifications that is most urgently needed at this time given the evolutionary process at play in the GNSS area.

The basic assumption here is that there are an increasing number of GNSS systems and satellites coming into play with enough interoperability to allow someone to depend on a multisystem approach to positioning, whereby they can cut signals arriving at small angles to the horizon and rely only on overhead satellites to determine their position.

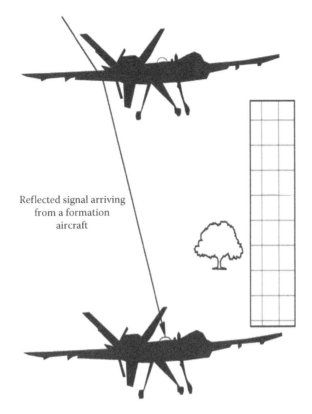

FIGURE 6.15
Antenna receiving high angle multipath for another formation aircraft.

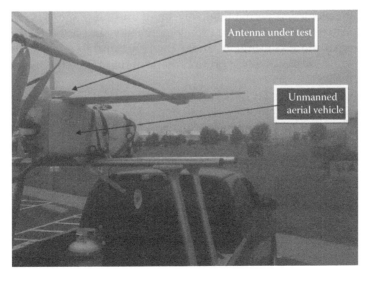

FIGURE 6.16
(See color insert.) A possible antenna test mount for urban canyon environment.

FIGURE 6.17
(See color insert.) A KSU Salina UAS asset. (Aerosonde Mark 4.7 UAS platform.)

Another key assumption is that by cutting off signals at angles low to the horizon one can reduce multipath errors. Both these assumptions lead to this new design criterion that, in terms of antenna speak, can be interpreted as having radiation patterns that have lowest gains at or near the horizon. Figure 6.7 illustrates how well the conical ring antenna performs in this area (it should go without saying that the old design criteria of good axial ratio, impedance matching, bandwidth, high gain in desirable directions still remain intact). Table 6.4 shows how well the conical ring performs at elevation angles close to the horizon and outperforms both the RSW and choke ring antennas by the new criteria.

The long-standing mechanical criteria of being a compact lightweight structure cannot be overemphasized in this application. Here also the conical ring betters RSW and choke ring in terms of size and weight. Figures 6.8 and 6.9 are significant in that it shows the RMS electric and magnetic field distribution is tightly held in the center. This is a likely reason why good roll off is achieved with a relatively small ground plane. Also from Figures 6.8 and 6.9 it is interesting to note how the FSS elements have impacted the RMS field distribution. It seems that the FSS elements under the cone tend to spread the field distribution outward toward the edges of the ground plane and are therefore more prone to edge effects.

Much of the discussion thus far has centered on the conical antenna's role as an antenna on a UAV. Given how the antenna rejects ray paths low to the horizon, it should provide some protection from jamming and spoofing sources that are ground based as well. It should be stated that no significant research has been done by the author with regards to protection from spoofing and jamming by the conical ring and that more investigation is needed.

Finally, studies are needed to see if the physical dimensions of the antenna can be reduced further without unacceptable loss in performance. The interesting problem of what happens when multipath rays actually enter the cone through near vertical angles from the body of the same aircraft as well as other aircraft flying in close formation is a real-world problem needing a solution. Finally, a real-world test where the antenna is taken for ride in an urban canyon hooked up to a multisystem receiver is important for gathering performance feedback.

Bibliography

Basilio, L., J. Williams, D. Jackson, and M. Khayat. 2005. A comparative study of new GPS reduced-surface-wave antenna. *IEEE Antennas and Wireless Propagation Letters* 4: 233–236.

Bayatpur, F. and K. Sarabandi. 2007. Single-layer, high-order, miniaturized-element frequency selective surfaces. *IEEE Transactions on Microwave Theory and Techniques* 55(5): 1239–1245.

Bayatpur, F. 2009. Metamaterial-inspired frequency-selective surfaces. PhD, University of Michigan, Ann Arbor, MI.

Bhattacharyya, A.K. 1990. Effects of finite ground plane on the radiation characteristics of a circular patch antenna. *IEEE Transactions on Antennas and Propagation* 38(2): 152–159.

Heng, L., T. Walter, and G. Gao. 2014. GNSS multipath and jamming mitigation using high-mask-angle antennas. *IEEE Transactions on Intelligent Transportation Systems* 16: 741–750.

Jahn, A., H. Bischl, and G. Heiβ. 1996. Channel characterisation for spread spectrum satellite communications. *Proceedings of IEEE Fourth International Symposium on Spread Spectrum Techniques and Applications (ISSSTA'96)*, Mainz, Germany, pp. 1221–1226.

Khan, S. 2014. A comparative analysis of the impact of different structures on multipath mitigation in the GPS context. *2014 IEEE International Symposium on Antennas and Propagation and USNC-URSI Radio Meeting*, Memphis, TN.

Kos, T., I. Markezic, and J. Pokrajic. 2010. Effects of multipath reception on GPS positioning performance. *2010 Proceedings ELMAR*, Zadar, Croatia, pp. 399–402.

Nedic, S. 2009. On GPS signal multipath modeling in dynamic environment. *2009 Aerospace Conference*, Big Sky, MT.

Rahmat-Samii, Y. 2003. The marvels of electromagnetic band gap (EBG) structures: Novel microwave and optical applications. *Proceedings of the 2003 SBMO/IEEE MTT-S International* 1: 265–275.

Rojas, R.G., D. Colak, and W.D. Burnside. 1995. Synthesis of tapered resistive ground plane for resistive antenna. *1995 Antennas and Propagation Society International Symposium*, Newport Beach, CA.

Schmid, A. and A. Neubauer. 2004. Comparisons of sensitivity limits for GPS and Galileo receivers in multipath scenarios. *Position Location and Navigation Symposium*, Monterey, CA, pp. 503–509.

Sciré-Scappuzzo, F. and Sergey Makarov. 2009. A low-multipath wideband GPS antenna with cutoff or non-cutoff corrugated ground plane. *IEEE Transactions on Antennas and Propagation* 57(1): 33–45.

Sievenpiper, D., L. Zhang, R.F.J. Broas, N. Alexópolous, and E. Yablonovitch. 1999. High-impedance electromagnetic surfaces with a forbidden frequency band. *IEEE Transactions on Microwave Theory and Techniques* 47(11): 2059–2073.

U.S. Air Force. 2004. 2004 Agreement on the promotion, provision and use of Galileo and GPS satellite-based navigation systems and related applications. Accessed November 9, 2014. http://www.gps.gov/policy/cooperation/europe/2004/gps-galileo-agreement.pdf.

U.S. Air Force. 2006. 2006 Joint statement on GPS-Galileo signal optimization. Accessed November 9, 2014. http://www.gps.gov/policy/cooperation/europe/2006/joint-statement/.

U.S. Air Force. 2007a. 2007 Joint fact sheet on GPS-Galileo cooperation. Accessed November 9, 2014. http://www.gps.gov/policy/cooperation/europe/2007/gps-galileo-fact-sheet.pdf.

U.S. Air Force. 2007b. 2007 Joint press release on civil signal design. Accessed November 9, 2014. http://www.gps.gov/policy/cooperation/europe/2007/MBOC-agreement/.

U.S. Air Force. 2007c. 2007 Joint statement on GPS-Galileo working group B. Accessed November 9, 2014. http://www.gps.gov/policy/cooperation/europe/2007/working-group-b/.

U.S. Air Force. 2008. 2008 GPS-Galileo working group B meeting. Accessed November 9, 2014. http://www.gps.gov/policy/cooperation/europe/2008/working-group-b/.

U.S. Air Force. 2010. 2010 Joint statement on combined performance of GPS and Galileo. Accessed November 9, 2014. http://www.gps.gov/policy/cooperation/europe/2008/joint-statement/.

U.S. Air Force. 2013. 2013 Interim report of ARAIM technical subgroup. Accessed November 9, 2014. http://www.gps.gov/policy/cooperation/europe/2008/joint-statement/.

U.S. Air Force. 2014a. New civil signals. Accessed November 9, 2014. http://www.gps.gov/systems/gps/modernization/civilsignals/#L1C.

U.S. Air Force. 2014b. Official U.S. government information about the global positioning system (GPS) and related topics. Accessed November 9, 2014. http://www.gps.gov/policy/cooperation/#europe.

Ward, P., J. Betz, and C. Hegarty. 2006. Interference, multipath, and scintillation. In *Understanding GPS Principles and Applications*, 2nd edn., E. Kaplan and C. Hegarty (eds.), pp. 279–295. Norwood, MA: Artech House.

Xie, P. and M. Petovello. 2015. Measuring GNSS multipath distributions in urban canyon environments. *IEEE Transactions on Instrumentation and Measurement* 64(2): 366–377.

7

Electromagnetic Interference to Aeronautical Telecommunications

Zhigang Liu

CONTENTS

7.1 Introduction

This chapter mainly analyzes the electromagnetic interference (EMI) of aviation communications equipment, introduces the EMI effects and response measures of aviation communications equipment, and then analyzes the protective measures of aircraft and ground stations against EMI; it also discusses EMI protection for aeronautical communications equipment.

7.2 Concepts of Aeronautical Communication Electromagnetic Interference

In the world, electromagnetic waves are everywhere. A variety of electronic devices such as radio, broadcast television, communication transmitters, and other radar, navigation equipment radiate electromagnetic energy into space during operation. This electromagnetic energy is intentional radiation, because it is consciously made to transfer a variety of data.

However, many electronic devices will also radiate electromagnetic energy into space during operation, such as from switching circuits, automatic ignition systems, and automatic control equipment. This electromagnetic energy is not deliberately manufactured, as it automatically appears in the circuit work process. They may affect other electronic devices, work and are undesirable, so is unintentional radiation.

Thus, because the electromagnetic environment (EME) is intentionally and unintentionally produced, it cannot be eliminated. When the intensity of electromagnetic radiation rises to a certain extent, it affects the normal operation of other electronic equipment and electronic systems. All aeronautical communication devices, in fact, must adopt appropriate measures to protect against electromagnetic radiation.

7.2.1 Related Terms

7.2.1.1 Electromagnetic Fields (EMFs)

Electromagnetic fields are formed by electric and magnetic fields. Electric fields refer to electric charges exerted on other electric charges. Magnetic fields refer to the magnetic object or moving electric charge effect on other magnetic materials and electric charges. EMFs take place throughout the electromagnetic spectrum; they are generated by natural or human activity. Some EMFs occur naturally, such as Earth's magnetic field, static electricity, and lightning. Also, others are established by the transmission and distribution of electricity, used in household electric appliances, communication systems, industrial processes, and scientific research.

The whole frequency spectrum is covered by natural and human-generated EMFs. *DC* (direct current) EMFs are EMFs that are nearly constant in time, and *AC* (alternating current) EMFs are those that vary in time. AC EMFs are characterized by their frequency range. The lower limit of 3–3000 Hz is defined as extremely low-frequency magnetic fields. Radio communications devices are operating at much higher frequencies, which are often in the range of 500,000 Hz (500 kHz)–3 billion Hz (3 GHz). The radio frequency (RF) sources of EMF typically include cellular telephone towers, broadcast towers for radio and television, airport navigation, radar, and communication systems. The high-frequency (HF) communication systems and very-high-frequency (VHF) communication systems are used by emergency medical technicians, utilities, and governments. Examples of local wireless systems are WiFi, cordless telephone, and so on.

7.2.1.2 Electromagnetic Interference

EMI is a repeated interruption or serious degradation of licensed radio communications service by any signal radiated in free space or conducted along signal leads, which disrupts the function of radio navigation or other safety services. It occurs when the EMFs produced affect operation of an electrical magnetic source adversely or other electromagnetic device. EMI can be caused by the source that intentionally radiates EMFs or one that does so incidentally. EMFs can be described in terms of the frequency or the times the EMFs changed direction in space each second. Radio communications services are of a great variety, such as AM/FM commercial broadcast, television, telephone services, navigation services, radar, and air traffic control (ATC). The licensed radio services together with some unlicensed radio services, such as WLAN or Bluetooth, may interrupt the EME.

Any EMI problem has three basic elements: interference source, path, and victim as in Figure 7.1. This chapter focuses upon the EMI situation of the airplane communication system. Because many interference sources are normally operated by battery power when used on board airplanes, the path is primarily radiated, rather than conducted. The limitations of such devices such as physical size and transmit power are given; the primary concern of EMI is for aircraft communication systems (victim).

7.2.1.3 Electromagnetic Compatibility

Electromagnetic compatibility (EMC) is a concept wherein the equipment and systems work properly in their EME and meet the requirements to run and at the same time do not produce EMI in the electromagnetic disturbance. The definition contains two aspects: First, the device should be able to work under certain normal EME and that the device should have a certain electromagnetic immunity (EMS). Second, electromagnetic disturbance, which is generated by the device itself, should not have too much impact on other electronic products, that is, EMI.

FIGURE 7.1
Three basic EMI elements.

In order to regulate the EMC of electronic products, all developed countries and some developing countries have developed standards for EMC. EMC standards guarantee that products can work in the actual EME of the basic requirements. It is called the basic requirements, that is to say, even though the products meet the EMC standards, in actual interference, problems may occur. Most national standards are based on the standards developed by the International Electrotechnical Commission.

7.2.1.4 High-Intensity Radiated Fields (HIRFs)

HIRF is higher radiation energy per unit area of the electromagnetic radiation, which is determined by the electric field and magnetic field strength of the electromagnetic wave. HIRF is radiation from the ground, ships, offshore platforms, or aircraft on radar, and radio, television, satellite uplink data, high-power transmitters. It is an EME problem caused by human activity, which is characterized by frequency bandwidth and long duration of action. In aeronautical communication systems, many radio antennas are in HIRF areas. The antenna is a radiator that converts the existing voltage–current form within the transmitter to a free-space electromagnetic wave, so it can help accomplish aircraft communications and navigation tasks. However, HIRFs are harmful to communication systems and some accidents are caused by HIRF, which has become an important factor affecting the safety of aircraft.[1]

7.2.1.5 Lightning and Lightning Strike

Lightning is a discharge of atmospheric phenomena, mostly formed in cumulonimbus. Cumulonimbus with changes in temperature and airflow will continuously move, and as the movement continues and friction powers, charged clouds are formed. Some clouds are positive charged, while others are negative charged. In addition, buildings and trees below the clouds accumulate charges due to electrostatic induction. With the accumulation of electric charge, the voltage of a thundercloud gradually increases, and when the thundercloud and the projections, ground with different electric charges, are adjacent to a certain degree, that is, when the electric field exceeds 25–30 kV/cm, intense discharge will occur, which is lightning, a simultaneously strong flash. Since the discharge temperature is high, there is a rapid expansion of heated air followed by a roar of explosion, which is thunder. In the lightning discharge process, objects that conduct lightning, that is, those being struck by it, would be destroyed. Also, an aeronautical communications system is affected by lightning transient EMI; hence, the design and operation of system should take into account the impact of lightning.

7.2.2 Civil Aviation Communication System Model

Civil aviation communication services mainly include ground communications, mobile communications in airport, telephone communications, ground–air communication, flight operation management communication, on-air radio, and confidential communications. The horizontal communications network is currently based on data communications network such as X.25 FR public network, frame relay, and asynchronous transfer mode ATM. Ground–air communication networks mainly include HF communications, VHF communications, ultra-HF (UHF) communications, and satellite communications.

The aeronautical communications system includes the VHF communication systems, HF communications systems, selective call system (SELCAL), intercom system, automatic direction finder (ADF), VHF omnidirectional range (VOR), instrument landing system (ILS), radio altimeter (RA), distance measuring equipment (DME), and ATC transponder and weather radar (WXR). The VHF system uses 118–135.975 MHz bands, channel spacing of 25 kHz, which is mainly for aircraft during takeoff and landing and for aircraft in controlled airspace during a two-way voice communication between the ground traffic control personnel. VHF signals use only linear wave propagation or line-of-sight propagation. The HF system uses 2–30 MHz bands, channel spacing of 1 kHz, which is mainly for aircraft communication with the airport during remote flight. The purpose of remote communication via reflection is to extend the communication distance up to thousands of kilometers between the ionosphere and the ground surface. SECAL is not an independent communications system, which coordinates with VHF and HF systems. The intercom system is used mainly for calls, broadcast, maintenance personnel calls, as well as audiovisual signals to travelers. ADF uses the long wavelength of 100–2000 kHz, which is for measuring ground navigation station with respect to the longitudinal axis of the aircraft position and for guiding the aircraft to fly. VOR uses the frequency band of 108–118 MHz, channel spacing of 50 kHz, to measure the VOR azimuth that determines the deviation of the aircraft relative to the selected VOR route. The ILS is the important equipment to ensure the safety landing of aircraft, by the localizer (LOC), glide slope (GS), and marker beacon (MB) components. The localizer uses 108.1–111.95 MHz bands, channel spacing of 50 kHz, which guide the aircraft to be aligned with the runway during landing. The glide slope uses 329.15–335.0 MHz bands, channel spacing of 150 kHz, which guide the aircraft fall along 3° angle to the runway during landing. The marker beacon uses 75 MHz bands, which check the correct aircraft height and speed through the beacon and provide the distance information of aircraft away from the runway threshold. RA uses 4200–4400 MHz bands, which are used to measure the actual height of the aircraft relative to the ground surface at the time of the approach and landing. DME uses 1025–1150 MHz and 962–1213 MHz bands, which are used to measure the distance of aircraft and DME beacon. ATC transponder uses 1030 and 1090 MHz bands, which are used to report the identification code and pressure altitude of aircraft to the ground control center and can be used to determine the horizontal position of the aircraft. WXR uses 9330–9400 MHz bands, which are used to detect typhoon target within the sector ahead of the aircraft and other obstacles, then select a safe route around the obstacle.

The principle of civil aviation communications systems is to specify powers, frequency, and antenna requirements, which ensure the reliable transmission of data across the wireless data link, while giving attention to the effects of propagation attenuation and noise effects relating to the free-space transmission, as illustrated in Figure 7.2.

The main link parameters contain the link operating frequency, the output power of transmitter, and the antenna parameters. In certain scenarios, increasing the output power levels of transmitter can reduce certain aspects of wireless EMI, but the method may

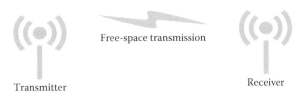

FIGURE 7.2
Civil aviation communication link.

cause more unnecessary EMI to other electronic and communication systems, so it is not practical for aviation communications devices with finite power sources.

The equations that can be used to calculate free-space transmission loss are given in the following. The distance variables are directly proportional. When the value of distance parameters is increased, the value of the free-space loss will increase at the same time. The frequency variables are the same with distance parameters to the value of the free-space loss:

$$L(\text{dB}) = 32.5 + 20\log_{10} d + 10\log_{10} f$$

where

  L means the free-space transmission loss in dB
  d means the free-space transmission distance in km
  f means the frequency in MHz

There are many factors that affect the propagation of radio waves, such as transmission power of transmitter, antenna gain, attenuation factors, and environmental noise. In addition, the ionospheric and tropospheric conditions of the environment also obviously take effect on the propagation of radio transmission in free space.

The spread of HF signals depends on sky wave propagation when their distances are approximately greater than 100 miles, depending on the ionization levels that exist in the ionosphere. In most cases, civil aviation HF communications propagate along the shortest path between the transmitter and receiver, and the signals propagated use the lower atmospheric region. The effects of the ionosphere are more obvious to the medium- and high-frequency regions signals than the VHF/UHF range signals, because it is refracted over a considerable vertical distance gradually. The highest frequency that is supported for reflection by the ionospheric layer is known as the maximum usable frequency.

The spread of VHF/UHF signals depends on ground wave propagation, which is almost always followed by line-of-sight paths. So it is required that there be no obstructions between the transmitter and receiver. The limitations of distance in this frequency range are primarily the functions of the antenna height of the transmitter and receiver.

7.2.3 Aeronautical Communications EMI Classification and Basic Measures

The International Telecommunication Union (ITU) is responsible for the allocation of frequency spectrum, and the frequency spectrum is assigned by state administrations, which are mostly government bodies. The assignments of each country are based on the allocations established in the ITU radio regulation. The assignment of frequency spectrum to airport and other aeronautical communication systems is the responsibility of the state civil aviation authority.

According to the allocation of ITU, aeronautical communication services are given primary status that is very important, so they must be protected from EMI from other communication services. Because broadcasting services and mobile communication services have secondary status, they must avoid causing EMI to any primary services and must be shut down if necessary.

EMI of aeronautical communication systems is primarily caused by undesirable voltages or current; they affect the performance of the aeronautical communication system and are processed as noise. Different EMI sources can be identified and mainly can be categorized into two kinds. One kind of EMI is from natural sources such as lightning, atmospheric effects, electrostatic discharge, sunspot activity, and reflections from the rough Earth surface. Other kinds of EMI are the man-made sources such as industrial activity, high tension electric cables, radar, and broadcasting transmitters.

The victims receive electromagnetic energy from the source of interference, and this energy has two main propagation mechanisms, radiation and conduction. EMI radiation is further divided into direct radiation and third-order intermodulation products radiation. The direct radiation originates from communication equipment and its interference uses the same frequency as the aeronautical communications system and radiates electromagnetic energy. The third-order intermodulation products radiation occurs when system operation creates one or more strong interference signals on certain frequencies and affects aeronautical communication systems. Similarly, intermodulation interference may also be generated in an aeronautical system receiver, which is caused by receiver being driven into nonlinearity leading to high-power broadcasting signals outside the band. When the interference is produced by two or more broadcasting signals that have a frequency relationship, an intermodulation product of the nonlinear process can occur within the RF channel through the aeronautical receiver.

Many techniques and measures have been taken in order to reduce the effects of EMI in communication systems. A variety of mitigation techniques have been used according to the interference nature, interference source, and interference mechanism of different communication services and systems. So not all mitigation technology measures are suitable for all cases. ITU provided a recommendation for EMI protection of safety services from undesirable emissions; many mitigation measures have been stipulated, which include the following:

1. Improving RF selectivity in order to reduce undesirable signals outside of the tuned bandwidth and provide good adjacent channel rejection performance. Beam down tilt so as to reduce interfering signal and also increase penetration.

2. Modification of antenna pattern that is much facilitated by corner reflectors and directional antennas, which when used direct a signal to the intended area of interest, thus minimizing interference to outside the service area.

3. Deployment of electronic RF filters such as notch filter and band-pass filters.

7.2.4 Aeronautical Communications Laws and Regulations

7.2.4.1 Aeronautical Communications EMI History

It can be said that civil aviation and radio communication have developed and grown up together. The development of radio communication technology has promoted the development of civil aviation communication technology. Since the 1920s, radio communication technologies have continually adapted to the ever-expanding need for aeronautical navigation and communication capability. At the same time, the need for reliable, secure communication, navigation, and surveillance (CNS) and air traffic management (ATM) for aeronautical communication has continually driven improvements in radio technology. Because civil aviation has become more dependent on radio communication for CNS, the EMI effects have become ever more critical and urgent. With very few exceptions, the aeronautical radio spectrum has been affected by other radio services, which have been coordinated and protected by law to minimize the potential of EMI. However, because

EMI have resulted in numerous aviation incidents and accidents over the years, which are effects of lightning, electrostatic discharge, and HIRFs from radars and broadcast transmitters, so EMI is carefully considered in all aspects of design now.

Since the inception of aviation, lightning has always been a problem, and flight missions need to occur in all weather conditions now. The National Advisory Committee for Aeronautics (NACA) established the Aircraft Safety Subcommittee in 1938. Lightning effects on aircraft have been studied by weather and lightning experts to determine what additional protective measures are needed. There is an influential history of airplanes struck by lightning, sometimes with catastrophic effects. Today, lightning strikes have minimal impact on the safety of airplanes, which has been well established by the FAA lightning protection regulations. The thorough technical guide to protecting aircraft from the effects of lightning has been provided by the textbook *Lightning Protection of Aircraft*.

In the early 1960s, compact radio receivers became available, which featured new transistor circuitry and were carried on airplanes by passengers. These devices could disrupt the ILS, VOR, and other navigation systems. In order to solve this concern, Special Committee 88 was formed by the Radio Technical Commission for Aeronautics (RTCA), which published *Interference to Aircraft Electronic Equipment from Devices Carried Aboard* (RTCA/DO-119) in February 1963. This began to control the use of compact radio receivers on board aircraft; this report recommended prohibiting the operation of portable radios during flight and set the stage for restrictions on passenger-carried compact radio receivers that are still in effect today.

By the 1970s, the problem of HIRF emanating from broadcast towers, radars, and point-to-point radio links arose; the new digital flight control systems needed to be robust against the effects of HIRF. The Department of Defense began to take into account for the control of EMI on aircraft systems in the United States. Many EMI-related handbooks relating to RF environment definition, procurement guidelines, hazards, grounding, bonding, and shielding exist. For commercial civil aviation, the RTCA worked with the FAA to set environmental standards and test procedures for aircraft equipment. Since the 1980s, NASA has documented several HIRF- and EMI-related events studies.

7.2.4.2 Laws and Regulations in China

With the development of China civil aviation, EMI to communication systems has become an urgent problem. The State Council and the Central Military Commission issued provisions on the protection of airport clearance in 1982, which specify the requirements of field emergency takeoff runway, airport, and clearance area between the airport runway and the highway. Then *Electromagnetic Environment Requirements for Aeronautical Radio Navigation Stations* (GB6364-1986) was enacted in May 1986 as the national standard in China.[2] This document analyzed the effect from nonaeronautical navigation services on all types of radio equipment, which include high-voltage power lines; electric railways; and industrial, scientific, and medical equipment, such as jamming caused by coupling the navigation station reflection or reradiation of the surrounding terrain; these factors may have harmful impacts on navigation information. This document lists the radio site protection requirements and measurement methods of many communication and navigation systems, which include nondirectional beacon (NDB), VHF/UHF DF, ILS, VOR, DME, TACAN, and PAR stations.

In August 1996, the Civil Aviation Administration of China (CAAC) released "Aeronautical radio navigation aids and ATC radar station sitting criteria" (MH/T 4003-1996). The industry standard standardized navigation, radar, and ATC radar terminal system setup location site environmental requirements, which is according to the civil aviation law, the GB6364-1986 national standard, and the International Civil Aviation Organization Annex 10 and other related articles of the convention on international civil aviation. It is related to terrain and houses, trees, embankments, and other active and passive objects, active and passive jamming, and the dual nature of overhead power lines, electric railways, and so on. In this document, the protections for radio interference are aimed at the various communications and navigation systems. It stipulates more detailed and specific requirements of ATC short-range and remote-range surveillance radar and ATC secondary radar stations. In 1997, the CAAC radio management committee gave notification of designated civil aviation airport EME protection areas, which reduce harmful interference to civil aviation radio stations.

In August 2010, "Electromagnetic environment protection requirements and measurement methods for VHF/UHF band radio monitoring station" (GB/T 25003-2010) was enacted as a national standard in China, which provides VHF/UHF band electromagnetic disturbance environment that allows values for radio stations and surrounding obstacles limiting requirements, provides for test method of EMI field strength, and defines the VHF/UHF band radio stations surrounding spacing requirements for minimum protection of high-power sources. The CAAC released "VHF air–ground communications ground system" (MH/T 4001.1-2006), which describes the VHF EME requirements. Likewise, "HF ground-to-air communications equipment general specifications" (MH/T 4002.1-1995) and "Electromagnetic environment requirements for short-wave radio receiving stations" (GB13617-92) were enacted, which describe the HF EME required of the receiving station. In December 2011, the ATC industry office of the CAAC released "Specification for civil aviation airport electromagnetic environment protection area delineation and protection requirements" (AC-118-TM-2011-01), which provides EME protection area delineation and protection requirements for civil aviation airport. In April 2013, the ATC industry office of the CAAC released "Civil airports and air traffic radio electromagnetic environment test specification" (AP-118-TM-2013-01), which standardized test civil airport and EME for air traffic radio stations.

With the rapid development of the civil aviation industry, the original GB6364-86 civil specifications have been unable to meet aviation EME protection requirements; the contradictions of urban construction and proliferation of navigation stations are increasingly prominent. Most of the standards currently cannot meet the development of civil aviation technology; the execution results are unsatisfactory. The rapid development of high-voltage power lines, power plants, roads, railways, radio communications has only increased radio interference. Also, in airport construction, EME issues have become increasingly prominent. The appropriate revision, "Electromagnetic environment requirements for aeronautical radio navigation stations" (GB6364-2013) was enacted in December 2013 as a national standard in China, which takes into account the development of military and civil aviation and EME requirements, taking into account the GB line with ICAO standards compliance. The practicality and operability of new standards is stronger and is conducive to EME.

HF/VHF EME protection requirements include VHF EME requirement and HF receiving station EME requirement.

7.2.4.2.1 VHF EME Requirement

The setting requirements for VHF aviation ground station EME are given in the following table:

Protect the signal electric field intensity (dBμV/m)				26
Electric field intensity	Protective rate (dB)		FM broadcasting	17
			Other	15
	FM broadcasting		Open value	−10
	Interference threshold (dBmW)		Cutoff value	−30

Protection distances of all kinds of interference sources for civil aviation VHF ground station are given in the following table:

Interference Sources		Protection Distance (m)
FM broadcasting	1 kW (include) less than	1000
	1 kW more than	6000
Electrified (electric) railways		300
Arterial highway	Level 2 and above highway	300
High-voltage transmission line	110–220 kV	300
	220–330 kV	250
	500 kV	300
Industry–science–medicinal RF equipment	Permissible variation meets the requirement	800

7.2.4.2.2 HF Receiving Station EME Requirement

HF receiving station EME must meet the MH 4002.1 and GB 13617-92 standards and EME requirements for shortwave radio receiving stations as given in the following table:

Interference Sources		Protection Distance of Different Levels of Stations (km)		
		First Level	Second Level	Third Level
Middle- and long-wave transmitting station (kW)	<50	10	7	3
	100–150	15	10	5
	>200	12	12	7
High-voltage transmission line (kV)	500	1.8	1.1	0.7
	200–330	1.3	0.8	0.6
	110	1.0	0.6	0.5
Shortwave transmitting station (directions in communications ¼ power angle) (kW)	0.5–5	4	2	1.5
	5–25	4–10	2–6	1.5–3
	25–125	10–20	6–10	3–5
	>120	>20	>10	>5
Shortwave transmitting station (directions in noncommunications ¼ power angle) kW	0.5–5	2	1	0.7
	5–25	2–5	1–3.5	0.77–1.5
	25–125	5–10	3.5–5	1.5–2.5
	>120	>10	>5	>2.5
Motor road	High-speed, first level	1	0.7	0.5
	Second level	0.8	0.5	0.3
Industry–science–medicinal RF equipment	Common	2	1.4	0.7
	More powerful	5	3.5	0.5

First-level short radio transmitting stations include the following:

1. Communication of central authorities, the state council, committee, frame, and the Chinese People's Liberation Army units that monitor operating department transmitting stations.
2. For defense, public security, and national safety and maritime security at stake and large building facilities or underground fortifications and trenches, massive antenna sites, and large high-gain antenna receiving stations.
3. Communication objects other than Asian countries, regional or distant-water fleets of receiving stations.
4. If a receiving station is located in the coherent region or urban areas, it must accord to the secondary-level station protection even though it is compliance with the aforementioned condition.

Second level shortwave radio receiving stations include the following:

1. Provinces, autonomous regions, and municipalities directly under the central receiving station, and People's Liberation Army–level unit directly under the receiving station
2. Asian countries such as Taiwan and its maritime communications.
3. The stations that meet these conditions include aeronautical communications and radar stations, navigation stations, and shortwave radio receiving stations. However, all three stations have equal amplitude reception.

Third-level shortwave radio receiving stations include receiving stations that were set by county or province, which include formal erection of antennas and other construction and which are responsible for more important communication tasks.[3]

Protection of the long-wave transmitter station spacing:

Transmit Power (kW)	Protection Spacing (km)		
	First-Level Station	Second-Level Station	Third-Level Station
<100	10	7	3
100–200	15	10	5
>200	20	12	7

Protection of the shortwave transmitter station spacing:

Transmit Power (kW)	Protection Spacing (km)		
	First-Level Station	Second-Level Station	Third-Level Station
0.5–5	4	2	1.5
5–25	4–10	2–6	1.5–3.0
25–120	10–20	6–10	3.0–5.0
>120	>20	>10	>5.0

Protection beyond the ¼ power angle of the directional antenna power:

Transmit Power (kW)	Protection Spacing (km)		
	First-Level Station	Second-Level Station	Third-Level Station
0.5–5	2	1.0	0.7
5–25	2–5	1.0–3.5	0.7–1.5
25–120	5–10	3.5–5.0	1.5–2.5
>120	>10	>5.0	>2.5

Shortwave radio receiving station (station) on the protection of high-voltage overhead transmission wire spacing:

Voltage Level (kV)	Protection Spacing (km)		
	First-Level Station	Second-Level Station	Third-Level Station
500	1.8	1.1	0.7
220–330	1.3	0.8	0.6
110	1.0	0.6	0.5

Road protection spacing:

Road Level	Protection Spacing (km)		
	First-Level Station	Second-Level Station	Third-Level Station
High-speed, first level	1.0	0.7	0.5
Second level	0.8	0.5	0.3

Protection pitch for industrial, scientific, and medical RF device users borders:

Industry, Science, Medicinal RF Device	Protection Spacing (km)		
	First-Level Station	Second-Level Station	Third-Level Station
Common	2.0	1.4	0.7
More powerful	5.0	3.5	1.5

Aeronautical radio frequency bands corresponding of radio stations:

Station Category			Radio Frequency Bands	Polarization
Communication		HF	2.8–22 MHz	Vertical
		VHF	118–137 MHz	Vertical
	Satellite Earth Station	C-band	3,968–3,991 MHz (Uplink)	—
		Ku-band	12,688–12,742 MHz (Downlink)	—

(Continued)

Station Category			Radio Frequency Bands	Polarization
Navigation	NDB		190–900 kHz	Vertical
	ILS	LOC	108–112 MHz	Horizontal
		GS	328.6–335.4 MHz	Horizontal
		MB	75 MHz	Horizontal
	DVOR		108–112 MHz	Horizontal
	DME		960–1215 MHz	Vertical
Surveillance	Primary radar	Remote	1,250–1,350 MHz	Linear
		Process	2,700–2,900 MHz	
	Secondary radar		1,029–1,031 MHz	Vertical
			1,087–1,093 MHz	
	ADS-B		1,089–1,091 MHz	Vertical
Meteorology	Boundary layer wind profiler radars		1,270–1,295 MHz	—
			1,300–1,375 MHz	
	Weather radar	S-band	2,700–2,900 MHz	—
		C-band	5,300–5,600 MHz	—
		X-band	9,300–9,700 MHz	—

Airport maximum permissible interference field strength and protection spacing requirements:

Station Category			Maximum Permissible Interference Field Strength
Communication	HF		20 dBμV/m
	VHF		9 dBμV/m
Navigation	NDB		22 dBμV/m (outside latitude 40°)
			27 dBμV/m (in latitude 40°)
	ILS	LOC	41 dBμV/m
		GS	12 dBμV/m
		MB	32 dBμV/m
	DVOR		19 dBμV/m
	DME		55 dBμV/m
Surveillance	Primary radar		−139 dBm
	Secondary radar		−106 dBm
	ADS-B		−85 dBm
Meteorology	Boundary layer wind profiler radars		−175 dBm
	Doppler weather radar	S-band	−151 dBm
		C-band	−150 dBm
		X-band	−150 dBm
	Common weather radar	C-band	−150 dBm
		X-band	−145 dBm
	Airport terminal area, Doppler weather radar	S-band	−153 dBm
		C-band	−159 dBm
		X-band	−159 dBm

Protection spacing requirements of communication equipment:

Station Category	Maximum Permissible Interference Field Strength		Protection Spacing Requirements		
	Sources of Interference	Allowed Values (dBμV/m)	Sources of Interference	Category	Protection Pitch (km)
HF	—	20	Medium-wave and long-wave transmitter station	<50 kW	10
				100–150 kW	15
				>200 kW	20
			Shortwave transmitter station (communication direction within 1/4 power angle)	0.5–5 kW	4
				5–25 kW	4–10
				25–120 kW	10–20
				>120 kW	>20
			Shortwave transmitter station (communication direction outer 1/4 power angle)	0.5–5 kW	2
				5–25 kW	2–5
				25–120 kW	5–10
				>120 kW	>10
			High-voltage transmission line	500 kV	1.8
				220–330 kV	1.3
				110 kV	1.0
			Automobile road	Highway	1.0
				Slow road	0.8
			Electrified railway		0.8
			Industrial, scientific, and medical radio frequency equipment	Common	2.0
				Multiple power	5.0
VHF	FM	9	FM	1 kW and less	1.0
				From 1 kW	6.0
			Electrified railway		0.3
			Automobile road		0.3
	Others	11	High-voltage transmission line	500 kV	0.3
				220–330 kV	0.25
				110 kV	0.2
			Industrial, scientific, and medical radio frequency equipment		0.8

Protection spacing requirements of navigation equipment:

Station Category		Maximum Permissible Interference Field Strength (dBμV/m)		
		FM	Industrial, Scientific, and Medical Radio Frequency Equipment	Others
NDB	Outside latitude 40°	—	28	22
	In latitude 40°	—	33	27
ILS	LOC	15	18	12
	GS	—	38	32
	MB	—	—	41
DVOR		22	25	19
DME		—	—	55

Protection spacing requirements of surveillance equipment:

Station Category	Maximum Permissible Interference		Protection Pitch			Remarks
	Sources of Interference	Allowed Values (dBm)	Sources of Interference	Category	Protection Pitch (km)	
Primary radar	High-voltage overhead transmission lines	−139	High-voltage overhead transmission lines	500 kV	1.0	
				220–330 kV	0.8	
				110 kV	0.7	
	Transformer substation	−139	High-voltage transformer substation	500 kV	1.2	
				220–330 kV	0.8	
				110 kV	0.7	
	Electrified railway	−136	Electrified railway		0.7	
			Nonelectrified railway		0.5	
	Automobile road	−136	Automobile road		0.7	
	HF heat sealing machine	−136	HF heat sealing machine		1.2	Counting from the plant
	HF furnace	−134	HF furnace	≤100 kW	0.5	Counting from the plant
	Industrial welding	−134	Industrial welding	≤10 kW	0.5	
	HF therapy machine	−134	HF therapy machine	≤1 kW	1.0	Counting from the workplace
			Farm power equipment	≤1 kW	0.5	
Secondary radar	—	−106	Same with the primary radar			
ADS-B	—	−85	Same with the primary radar			

Protection spacing requirements of meteorology equipment:

Station Category		Maximum Permissible Interference (dBm)
Doppler weather radar	X-band	−150
	C-band	−150
	S-band	−151
Common weather radar	X-band	−145
	C-band	−150
Airport terminal area, Doppler weather radar	X-band	−159
	C-band	−159
	S-band	−153
Boundary layer wind profiler radars		−175

7.3 Control of Aeronautical Communications EMI

7.3.1 Effects of Aeronautical Communications EMI

7.3.1.1 EMI Impact on Aeronautical Communications

Each airport has its own set of primary and standby frequencies for communication in the VHF AM, aircraft, and from various airports, while aircraft must frequently change frequency in deferent airport stations, so that the bandwidth of the antenna must be a full band. Aircraft design must limit the shape and number of antennas so as to reduce drag and conform with the principles of aerodynamics. A full-band antenna can be used as much as possible; generally this reduces antenna performance related to signal attenuation of outside frequencies when compared with the single-frequency antenna. In the practical application, VHF radio communication uses the same vertically polarized wave, which cannot reduce co-channel interference when compared with directional antenna VOR/ILS communication systems. An AM receiver can choose FM frequency signal with a linear transformation as with processing for AM signal, and when frequency band is suitable, the AM signal will be demodulated to sufficient strength.

VOR and ILS receivers use horizontal dipole antenna and horizontal polarization wave, so there is a strong directionality. Polarization properties of the antenna determine certain attenuation to ground-based communication signals with vertical polarization. The low-frequency demodulation processing circuit of VOR and ILS receivers processed 30, 90, or 150 Hz fixed frequency signals, so it is easy to filter out conventional 300–3000 Hz voice signal. The marker beacon receiver used a fixed transducer capacitance to match the shortened horizontally polarized antenna. This measure does not only have a vertical polarization signal damping effect to the communication systems, but also the sensitivity of the receiver is lower and the gain of the interference signal is smaller. This unique working way of VOR and ILS navigation system determines that they have strong antijamming capability, thus avoiding interference.

EMI directly affects aeronautical communication and navigation systems; the conclusion can be drawn by observing anomalous aircraft system behaviors that may result from EMI. A general description of EMI effects for several aeronautical communication and navigation systems observed during the EMI signal is provided there, and operational methods to handle these effects are explored.

The effects on VHF communication system are varied according to specific modulation types of EMI signal, as well as the particular aircraft radio model. Some VHF systems were silent without any alerting to the crew or any interference indication prior to reaching the threshold suddenly. Other VHF systems were subject to distortion and undesirable noise when the power level of the interfering signal was increased, and they were judged to be unusable by the pilot for voice communication. Without specific failure indications or warnings during EMI testing, signal occurred for the VHF system. Failure of the VHF system occurred when the EMI signal was transmitted from numerous locations in the passenger cabin, and their manner and level was similar. Thus, if EMI is occurring in situations of undesirable noise and selection of an alternate VHF communication system, it would not likely mitigate the effect of EMI.

For ILS (localizer and glide slope) and VOR, the EMI signal can cause visible variations in both vertical and horizontal indicators. There are two types of variations: offsets and fluctuations of the course and glide slope. The navigation display has the indicator of declaring *operational EMI failure*. But if this indicator has been indicated by a failure flag, it may show that either the reference signal was too strong or the received signal was too weak. VOR has an indicator of *system EMI failure*, thus the localizer or glide slope failure flag appears on the navigation display.

The most common effect of EMI on DME was on displayed DME data. These effects were generally more serious on one particular aircraft system, therefore, operational procedures may be helpful for mitigating EMI impact. On some using live DME signals systems, the indicated distance to that site was varied up to 1 nautical mile when an EMI signal was present. The Morse code identifiers were affected sometimes. Therefore, the Morse code identifiers cannot be verified if the indicated direction was disturbed.

For the ATC transponder, operational EMI failure is defined as a drop below 90% in aircraft replies to interrogations. It may be helpful for operational procedures to include checks of the ATC reply indicator if EMI is suspected. EMI effects to ATC receivers gradually increased with higher EMI signal power levels. Thus, EMI is more likely to affect system processing of ATC replies from distant airplanes rather than nearby airplanes.[4]

The allowed values are shown in the following table, which are aeronautical communications systems for industrial, scientific, and medical equipment interference and attenuation characteristics.

Protection Target	Frequency Range (MHz)	Protection Rate (dB)	Interference Attenuation Rate	Interference Allowed Values (dBμV/m)
NDB	0.15–1.75	9	$d^{-2.8}$	85
LOC/GS/VOR	108–400	14	d^{-1}	40

Using the following formula, we can calculate the interference protection zone for industrial, scientific, and medical equipment:

$$d = 30 \times 10^{\left(\frac{E_{30} - E_S + R}{20A} \right)}$$

where

    d means the protection distance, that is, industrial, scientific, medical equipment from the ground or airborne receiving equipment

    E_{30} means interference allowed values of industrial, scientific, and medical equipment

    E_S means signal strength of protection target

    R means protection rate

    A means protection against interference in a decaying exponential decay rate

The signal strength of radio equipment is calculated as follows:

$$E_X = E_S + 20 \lg \left(\frac{d_s}{d_X} \right)$$

where

    E_X means signal propagation field strength

    E_S means the known signal strength

    d_X means the propagation distance

    d_s means the distance from the signal source

7.3.2 Common Interference Analysis

The basic types of aeronautical communications and navigation frequency interference are intermodulation interference, right-of-band interference, and co-channel interference.

Intermodulation interference is divided into the transmitter intermodulation interference and receiver intermodulation interference.

Transmitter intermodulation interference occurs when the signals of multiple transmitters fall in the bands of another transmitter. In the final amplifier nonlinear interaction, unnecessary combined frequency of the received signal appears, which causes interference to the received signal with the same frequency as the combination.

The transmitter intermodulation interference can be reduced by increasing the transmitter coupling loss (antenna isolation). We can take the following measures:

1. Increase the distance between the transmitter antennas. In order to meet the isolation >50 dB, the transmitter antenna should be placed vertically, spaced apart about 3λ, or horizontally, spaced apart about 40λ.

2. Unidirectional isolator. Inserting high-Q band-pass filter between the transmitter output and feeder, increasing the frequency of isolation.

3. Try to avoid intermodulation frequency group assignments in tall buildings, mountains, and tower positions.

The intermodulation interference of the receiver refers to intermodulation frequencies along with the single or multiple strong signals simultaneously entering the receiver. The interference occurs in the front-end nonlinear circuits of the receiver (e.g., HF amplifier, mixer). If the intermodulation frequencies fall into the receiver intermediate frequency band, interference occurs.

The most common approach is to add a high-Q value of the resonator filter and improve the selectivity of the front of input circuit, to suppress the signal outside the receiver frequency.

Right-of-band interference is generated by the stray radiation of transmitter and spurious response of receiver interference. At low frequencies, parts of VHF and UHF mobile communications equipment use the crystal oscillator to obtain a higher-frequency stability, particularly the base station transmitter. The desired transmit frequency f_t can be obtained by repeated octave of main vibration frequency f_0. Because of the nonlinear and multiplier amplifier effect, this results in a large number of harmonic frequencies. If the frequency characteristic of filter circuit is poor, these harmonics will be amplified and radiated together with f_t, to interference operation of the receiver in the corresponding frequency. The interference can be generated on one or more frequencies near the outside of the transmitter bandwidth. In short, the interference is due to the value of the transmitter spurious radiation caused. So all types of transmitter spurious radiation values have stringent requirements in national standards. According to the provisions of the National Transmitter Technical Specifications: when the qualified transmitter carrier power is greater than 25 W, the radiated power of discrete frequencies should be less 70 dB than the transmitter carrier power. If the transmitter spurious radiation is too high, it usually introduces many problems, such as too much multioctave frequency, poor selection of multiplier output circuit, poor shield isolation between the multiplier and other factors, and so on.

In addition to the useful signals that were received by the receiver, other unwanted signals were received at other frequencies. The kind of response capabilities of other unwanted signals, commonly referred to as spurious response, is related with the frequency purity of the receiver local oscillator. Spurious response of the superheterodyne receiver was mainly IF and mirror-frequency response. The IF response is caused when the interfering signal frequency is equal to the intermediate frequency of the receiver, the interference signal from the receiver input circuit, then leaks into the HF amplifier circuit; thus, its suppression is not enough, so IF signals go directly to the circuit. Mirror-frequency response is caused by subtracting local oscillator frequency from the image frequency, through the IF loop. Therefore, the receiver can respond in the image frequency. Mirror-frequency interference has occurred in military and civil aviation communications and navigation frequencies. Because aviation communications and navigation radio receiver selectivity standards are poor (spurious response restrain, adjacent channel selectivity) and cause mirror interference easily, installation of narrowband filter at the input of the receiver is the common way to handle mirror-frequency interference.

Co-channel interference occurs when its frequency is the same or close to the interference frequency and the desired signal frequency and when the signal is received in a similar manner.

7.3.3 Common Sources of Interference and Countermeasures

7.3.3.1 Analysis of Interference Sources

With the application of a wide variety of radio and nonradio equipment in recent years, the interference of aviation communications and navigation equipment has increased.

7.3.3.1.1 Radio Equipment

1. *High-power cordless phones*: High-power cordless phones can interfere with aviation frequencies, through co-channel interference. First, their transmit power is too large, resulting in spurious radiation falling into the frequency spectrum dedicated to aeronautical communication. Second, direct use of such telephones will result in emission of frequency dedicated to aeronautical communications and navigation. Illegal setup and use of high-power cordless telephones poses greater risk, because it will cause random interference effects.

2. *FM radio/TV*: First, the 88–108 MHz band of FM radio is adjacent to the dedicated 108–137 MHz band of aeronautical communication and navigation system. Second, FM radio transmitter power is too stong, and numerous broadcast antennas are located on the same mountain and towers, so it is easy to generate intermodulation signals. Third, the cable additions channel uses the dedicated frequency of aviation communications and navigation system directly. There are multiple cable links, trunk amplifiers, cable amplifiers, and splitters, in TV lines. A bad cable connection, damaged shielding layer, or cable signal leakage that reaches a certain intensity will cause harmful interference to the dedicated frequency of aeronautical communications and navigation.

3. *Illegal radio transmitter*: First, the power of an illegal radio transmitter is too strong. Second, its frequency assignment is illegal, and its station set is irrational. Signals of different frequencies are transmitted simultaneously to form intermodulation products that fall within the aeronautical communications band. Third, an aging transmitter equipment usually fails, and decreased work performance indicators, mainly spurious emission indicators, fail, causing the right-of-band radiation to affect aircraft communications.

4. *Mobile phone base stations and repeaters*: A frequency of 800 MHz or more is generally close to the harmonic frequencies of airport navigation devices, so intermodulation and other spurious signals possibly cause interference to airport navigation devices.

5. *Aviation devices*: Many radio components are used in aviation, and they are prone to interference. There are main causes of interference: first, equipment failure or spurious emissions and second, equipment installation configuration is not standardized, resulting in emission power shortage, insufficient sensitivity, or multiple devices affecting each other.

In addition, with the rapid development of wireless technology and the increasingly wide range of wireless applications, some new sources of radio interference will also appear.

7.3.3.1.2 Nonradio Equipment

Nonradio equipment, found in industry, science, and medicine, will emit a radio signal during operation, and the RF signal is generally wide, and its signal strength large. But these signals are not fed through amplification systems, and therefore, the attenuation is obvious to disappear after the propagation distance of several hundred meters. In theory, they do not often interfere with the aircraft, but if they are at a short distance facing empty stations, they will interfere with aviation communications and navigation frequencies.

As the high-voltage power transmission line voltage level continues to increase, the wire surface corona discharge and other opportunities also increase, and the effect of corona discharge produces radio interference. The essence of radio interference is corona discharge process, and it produces some harmful electromagnetic waves of relatively wide frequency from the low 50 Hz to an HF range. These frequencies can interfere with normal operation of surrounding radio communication devices.

Radio interference of the transmission line is mainly caused by the wire corona discharge; the surface contamination of the insulator also causes spark gap, clamp corona, spark discharge, and glitches discharge. The radio interference of an electrical substation is caused by many factors and is complex. It will have a major spark discharge switch operation from the substation at normal operating and fault conditions. It also causes partial discharge due to damage, contamination, and other reasons such as corona discharge of bus and device, etc.

Transmission line radio interference includes uniform interference, nonuniform interference, and pulse interference; in theory, any radio device will be affected by frequency interference, in fact, mostly, AM radio, communications (0.5–12 MHz), and television interference.

The degree of interference to radio transmission mainly depends on the distance from the transmitter device, performance of the communication device, the antenna orientation, and transmission lines of various parameters, such as wire arrangement, level voltage tower height, size sag, and weather conditions, such as sun, rain, or fog.

AC overhead transmission lines cause radio interference limits under normal circumstances, 110–500 kV high-voltage lines can generate a frequency of 0.15–30 MHz of radio interference. AC high-voltage overhead transmission lines characteristic of radio interference distance can be represented by the formula

$$E_X = E + k \cdot \log \frac{400 + (H - h)^2}{X^2 + (H - h)^2}$$

where

E_X means interference field strength
E means interference field strength, 20 m from the line
X means the distance from the line
H means the line height from ground
h means the measuring instrument height from ground
k means attenuation coefficient

7.3.3.2 Remedy

Because ionospheric interference must be reflected back to the ground, thus, the height and the electron density of the ionosphere changes with the seasons, day and night. Shortwave communications will be affected by distance and signal strength. In order to ensure stable and reliable shortwave communications, the appropriate frequency should be selected based on factors such as seasons, day and night, and area. Generally, a higher frequency should be selected in summer and daytime for appropriate communication; a lower frequency should be suitable for winter and night.

To reduce the effect of the dead zone, one may increase the power of the ground station to increase the wave propagation distance. The operating frequency can be reduced to shorten the distance between the nearest sky wave reflections, thereby narrowing the scope of the dead zone, and also improve flight altitude, because the higher the altitude, the smaller the dead zone range.

To increase the communication distance, the VHF antenna should be mounted higher. In air–ground communications, as aircraft fly higher, that is, the distance between the aircraft and the ground is greater, the farther the ultrashortwave propagation distance is.

Safe professional use of radio frequencies is the premise and foundation to ensure flight safety, so it must attach importance to radio management. In particular, airport communications departments should increase the protection of the EME around the airport, rely on local government and local radio management departments, and formulate relevant regulations through the legislative branch of government. Airport communications departments should strengthen communication and cooperation with the local radio management departments. If radio interference is found, one should first analyze the intensity, orientation, and type of the interference source and then seriously make a record and provide a basis for finding the source of interference. Second, one should appeal to the local radio management departments in a timely manner. One should actively seek the establishment of monitoring equipment at the airport by local radio regulatory authorities, so that aeronautical frequencies can be monitored to detect interfering signals. Industrial and commercial administrative departments should intensify efforts to investigate and ban the production and sale of illegal radio equipment.

Aeronautical radio equipment, being mobile, causes mutual interference unavoidably. The interference may be caused by improper operation of personnel or equipment, so authorities must take scientific analysis seriously, strengthen management in strict accordance with equipment operation and maintenance procedures, conduct a comprehensive inventory of radio frequencies and equipment, punish illegal use of radio equipment, and subject equipment for testing on a regular basis.[5]

7.4 Aeronautical Communications System Protection Requirements

7.4.1 Aircraft Design Protection Measures

As the complexity of electronic equipment and electrical systems increase, interference of airborne electronic equipment is becoming increasingly serious. Therefore, it has become a very important task to eliminate such interference, so that all airborne electronic equipment can be working in conditions of EMC. The main measures taken are as follows:

7.4.1.1 Grounding

For radio equipment, the aircraft fuselage structure serves as the ground, which is the common port. In order for the radio device to perform its function properly, it should maintain an appropriate balance between aircraft and antenna structure. That is, the surface potential of the aircraft should be stable. However, the operation surface of the aircraft is separated from the rest parts sometimes, so it causes difference by body surface and operation surface potential. If we do not alleviate the situation by grounding, it will affect the operation of radio equipment. Therefore, it is connected between the plane operating surface and the ground.

7.4.1.2 Overlap

An aircraft in flight will generate a lot of static friction between airframe and dust in the air, water, and other particles in flight. If there is poor contact between the parts of the airplane, the charge generated by each metal member will vary, so that there is a certain potential difference between the plane metal surface. When the potential difference reaches a certain value, it will cause an electrostatic arc discharge. Arcing occurs in the form of short pulses, which generate noise and will cause EMI in the radio spectrum to antennas and other radio equipment.

Overlap can provide a low-resistance path for all metal parts within the aircraft, eliminating the potential difference between the metal parts, thereby eliminating EMI caused by electrostatic discharge, but also maintain a constant flow of low-resistance path on the charge. It is used to connect devices that have lap belts and lap folders commonly. Overlap between the radio and the airplane structure provides a low-resistance path between the common port, which also reduces static interference.

7.4.1.3 Shielding

Shielding is one of the most effective methods of removing noise; its basic purpose is to shield the RF noise in electrical energy, which reduces the electromagnetic energy radiated outward. For example, the use of shielded antenna feeder cable can reduce the radiation of electromagnetic energy outside and can prevent external electromagnetic energy from entering into the feed line. The data bus on the aircraft also require mask processing and proper grounding.

In a radio device, those circuits that radiate HF electromagnetic energy are used together with the shield cover to prevent them from affecting other circuits. Electromagnetic radiation energy from the inner surface of the shield is conducted into the ground. In the case where the filter cannot be used, shielding is particularly effective. For example, when the electromagnetic radiation source radiates energy, the receiver input circuit and other various circuits that are connected will receive the noise. In this case, we should install filters on each affected line, which is virtually impossible, so it is best to take radiation shielding into effect, because it will bar the radiation source that is enclosed within the shield, thereby reducing the electromagnetic energy radiated outward.

For example, ignition and spark plugs on the aircraft are usually shielded to eliminate interference. Another example that will produce EMI between the motor and solenoid switch is the use shielding, but the amount of miss-out interference noise from the shield is quite big; thus switching between the motor and the installation of a magnetic filter is needed to further reduce noise. This may be just a simple filter capacitor and may also consist of capacitors and an HF choke. After the installation of filters, usually there is no need to take measures to shield between the motor and the solenoid switch.

7.4.1.4 Static Discharger

The purpose of the installation of a static discharger on the aircraft is to release static charge, which is generated during the flight on the body, reducing interference to radio equipment. A static discharge device is usually installed in the rear edge of the control surfaces, wing tips, and vertical stabilizer surface. It should be noted that a static discharger will not protect against lightning strikes.

7.4.1.5 Lightning Protection

Lightning protection has been implemented in the shape of the aircraft design. The external structure and enclosure of the aircraft should use almost all metal materials; otherwise, they will not have sufficient thickness to withstand lightning strikes, so the shape of the outer structure can be said to constitute the plane's basic lightning protection. The metal surface is a shield that prevents lightning from damaging the cabin of the aircraft, and also blocks the electromagnetic energy from entering the aircraft cable. Inside the aircraft, arresters are installed as some parts are also vulnerable to the effects of lightning, such as the alternator and antenna tuners of the aircraft. Once an aircraft is struck by lightning, it must undergo a comprehensive inspection.

7.4.2 Ground Station Design Protection Measures

7.4.2.1 Airport Requirements

The EME protection zone of a civil airport includes two parts: the regional civilian airport civil aviation radio (station) EME protection zone and civilian airfield EME protection zone.

EME protection regional of civil aviation radio (station) of a civil airport includes the rectangular range that is occupied by the civil airport runway. Both ends of the runway centerline and extension cords as a benchmark are extending 500 m, respectively, to the sides. EME in the civilian airport zone settings prohibits the unauthorized change of its technical parameters. Civil airport authorities and civil aviation radio station are set using units, which should establish civil airport EME protection zone patrol system. When new facilities or behavior that may affect airport EME (including changes in topography) are discovered, one should promptly report them to the local civil aviation authorities. In case of an emergency or special circumstances, one should report directly to the local radio regulatory agency.

7.4.2.2 NDB Station Requirements

NDB and ADF work together for the determination of the aircraft relative azimuth with navigation station and guide the aircraft along the intended route of flight, homing, approach, and landing, which include close NDB, remote NDB in airports, and route NDB. The close NDB and remote NDB are set in the runway center extended line; the distance is 900–11,100 m from the landing end of the runway. Route NDB is usually set below the route turning point and the checkpoint. Thus, the signal coverage area radius of the close NDB should be not less than 18.5 km, not more than 70 km; signal coverage area radius of the remote NDB and route NDB should be within a radius of 150 km. And the minimum signal strength should be more than 70 µV/m outside latitude 30° and should be more than 120 µV/m in latitude 30°. It is 9 dB for industrial, scientific, and medical equipment interference protection and is 15 dB for a variety of other active interference protection in NDB signal coverage area. The terrain should be flat, open in the radius range of 100 m from the center of NDB antenna. The minimum separation distance is allowed between the centers of NDB antenna and all kinds of terrain as shown in the following table. The communication and power cables near the NDB station should be buried, with a distance of 150 m from the center of the NDB antenna. There should be no obstruction outside 50 m in the bottom center of the antenna as a benchmark over a vertical opening angle of 3° as shown in Figure 7.3.

Terrain name	Minimum Separation Distance
Higher than 3 m of trees, buildings, and roads	50
Railways, overhead low-voltage power lines, communication cables, overhead high-voltage transmission lines below 110 kV	150
Hills, dams	300
Overhead high-voltage transmission lines above 110 kV	500

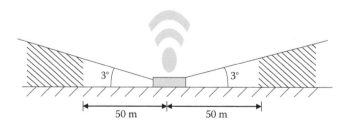

FIGURE 7.3
(See color insert.) No obstruction outside 50 m in the bottom center of the antenna as a benchmark over a vertical opening angle of 3°.

7.4.2.3 LOC Station Requirements

The localizer is an integral part of the ILS system and works in conjunction with the aircraft receiver, which provides the landing alignment guidance information of aircraft. Localizer antenna array is usually set in the extension line of the runway centerline, and the distance from the runway end is 180–600 m. The localizer transmitter sets the horizontally polarized sector synthetic field and the signal coverage area: the range of 10° of the runway centerline is 46.3 km, and the range of about 10°–35° is 31.5 km, as shown in Figure 7.4, and the minimum signal strength is 40 μV/m in the coverage area.

Because of terrain conditions, the localizer antenna cannot be set in the extended runway centerline; the offset settings can be used. Thus, the maximum allowable deviation deflection angle is 3°, and the distance from the runway centerline must be less than 160 m, and the bias settings are only limited as ILS of class I, as shown in the bias antenna array configuration in Figure 7.5.

In the signal coverage area of the localizer, the radio interference protection ratio is 17 dB to FM signal, is 14 dB to industrial, scientific, and medical equipment, and is 20 dB to a variety of other active interference.

The protected area of the localizer site is denoted by circular and rectangular regions, and the center of the circle is the localizer antenna; its radius is 75 m. The length of rectangle is 300 m to runway end (whichever is greater) from the localizer antenna extending away along runway centerline; the width is 120 m. If the localizer antenna has radiation characteristics of single direction, and radiation pattern around the field is better than 26 dB or more, the protected areas are not included in the shaded region in Figure 7.6.

The machine room of the localizer should be set within the range of the antenna array, which is less than 30° the direction of antenna array. According to the local topography, road and power conditions can be set on either side of the antenna; the distance between the antenna center and the room is 60–90 m.

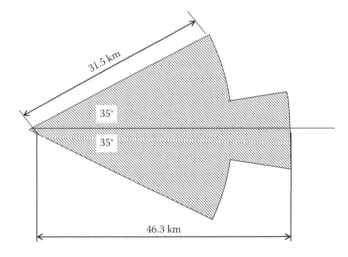

FIGURE 7.4
The localizer transmitter makes the horizontally polarized sector synthetic field and the signal coverage area: the range of 10° of the runway centerline is 46.3 km, and the range of about 10°–35° is 31.5 km.

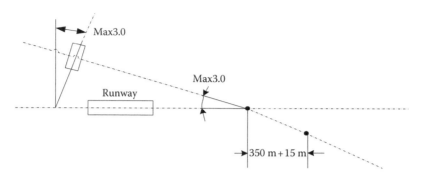

FIGURE 7.5
Bias antenna array configuration.

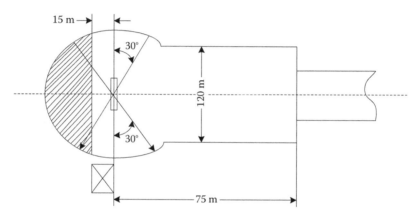

FIGURE 7.6
Site-protected area of the localizer.

There should be no obstacle in the protected areas of the localizer, and the beacon power cables and communication cables should be buried in the protected areas. Vehicles or aircraft should not be parked in protected areas, and there should be no ground transportation activities. Even if airfield lighting facilities must be set in the protection zone, the use of nonmetallic materials should be maximized, and its height and surface area should be as small as possible to ensure that its impact on radio navigation signals minimized. Thus, there should be no tall buildings, large metal reflectors, or high-voltage transmission lines in the area that is in the range of 15 m within the antenna forward range of 20° and within the distance of 3 km.

7.4.2.4 GS Station Requirements

The glide slope is an integral part of the ILS system and works in conjunction with the aircraft receiver too, which provides the landing vertical guidance information of aircraft. The glide slope antenna array is usually set within the landing end of the runway off to the side, which is of minimal impact to flight, and the distance from the runway center may be 75–200 m, in which 120 m is usually chosen. The distance of glide slope retreat away from the runway entrance is usually 200–400 m, which can be adjusted in accordance with the following parameters: glide angle, height of baseline data point, terrain slope, and so on.

The glide slope transmitter makes the horizontally polarized sector synthetic field and the signal coverage area: the antenna forward horizontal range of 8° and the antenna forward vertical range of 0.450°–1.750° (θ is the slip angle) is 18.5 km, as shown in Figure 7.7, and the minimum signal strength is 400 μV/m in the coverage area.

In the signal coverage area of the glide slope, the radio interference protection ratio is 14 dB to industrial, scientific, and medical equipment and is 20 dB for a variety of other active interference. The protected areas of the glide slope are shown in Figure 7.8. Region A should not have any obstructions, roads, or crops; the height of weeds should not exceed 0.3 m. The longitudinal slope of the region should reflect the runway slope, and the cross slope should be less than 1%, and the area must be flat with no less than 4 cm of height difference. In the region, there should be no parked vehicles, machinery, or aircraft and should not have ground transportation activities. The power cable and communication cable should be buried underground through the region. Region B is 600 m front of the glide slope antenna; there should be no railways, highways, dedicated surround channel of airports, buildings (except localizer room), high-voltage transmission lines, dams, trees, hills, or other obstacles. The height of the localizer room and the obstacle outside 600 m should not exceed the limits of the runway end clearance. The region should be flat ground; uneven terrain height allowable values are related with the distance to the glide slope antenna and antenna height and other factors, and the formula is as follows:

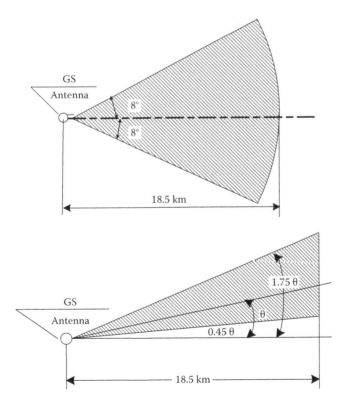

FIGURE 7.7
The antenna forward horizontal range of 8° and the antenna forward vertical range of 80.450°–1.750° (θ is the slip angle) is 18.5 km.

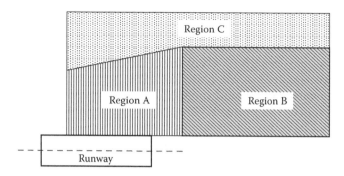

FIGURE 7.8
(See color insert.) The protected areas of the glide slope.

$$Z < 0.0117 \frac{D}{N}$$

where
   Z are the uneven terrain height allowable values
   D is the distance to the glide slope antenna
   N are the wavelengths number of sideband antenna height or the wavelength number of
      capture the effect middle antenna

Region C should not have rail and road presence (except dedicated surround channel of airports); there should be no higher buildings, high-voltage transmission lines, dams, trees, hills, and other obstacles than the airport side clearance restrictions; terrain slope in the region should not be more than 15%. The glide slope room should be set in the rear or side rear of the glide slope antenna, and the distance from the glide slope antenna is 2–3 m.

7.4.2.5 MB Station Requirements

A marker beacon (MB) station is part of the ILS; middle and inner marker beacon station, which is set in the extended runway center line in accordance with the requirements of the external and the operating frequency, is 75 MHz. MB emitted into the air vertically tapered synthetic field, and the signal coverage areas are the longitudinal width of 100–200 m in the height of 20–40 m, the longitudinal width of 200–400 m in the height of 60–80 m, and the longitudinal width of 400–800 m in the height of 350–600 m.

When MB is influence by the terrain, the radiation pattern will be distorted, causing deviation mark locations. When MB and NDB are colocated setting, its antenna is disposed in the extended runway center line, and from NDB antenna 10–30 m. When ground conditions do not permit the spot, MB antenna can also be mounted directly on the NDB room roof.

The active interference protection ratio of marker beacon signals is 23 dB within the coverage area. Outer MB is 6,500–11,100 m from the runway, in which 7,200 m is usually used, and middle MB is 1050 ± 150 m from the runway, and inner MB is 75–450 m from the runway. The outer MB and middle MB deviation should not exceed 75 m from the extended runway centerline, and the internal MB deviation should not exceed 30 m from the extended runway centerline.

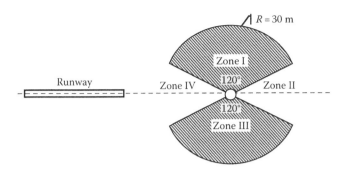

FIGURE 7.9
The protection zone I, II, III, and IV.

Within the protection zones I and III as in Figure 7.9, the benchmark for the ground grid or MB antenna lowest unit should not be exceeded. Moreover the standard units without directional radio antenna should be outside the room, and the distance from the MB station should more than 30 m, with a vertical opening angle of 20° for obstructions. And within the protection zones II and IV, there should not be exceeded according to network or MB antenna lowest unit as a benchmark, in addition to standard units without directional radio antenna outside the room, and within a distance marker beacon station 30 m, vertical 45° opening angle of the obstacle as in Figure 7.9.

7.4.2.6 DVOR Station Requirements

DVOR operates in the 108–117.975 MHz band and is in cooperation with the aircraft receiver and provides comprehensive guidance information to guide aircraft along a predetermined route (line) flight arrival and departure and approach for aircraft. DVOR has a reflective grid surrounding the station; the multipath interference and reemitted radiation to the radio signal can distort the radiation pattern, resulting in channel bend, swing, and jitter, which in turn affects flight safety.

The DVOR antenna base center is a reference point to the antenna reflector net plane for plane. There should not be any obstacles beyond the datum height within a radius of 100 m, and there should be no roads, buildings, dams, hills, trees, and other obstacles within a radius of 200 m that exceed the height datum. The opening angle relative to the plane should not exceed 1.5°. Horizontal aperture angle should not exceed 7° within the radius of 100–200 m vertical. Obstacle vertical opening angle relative to the reference plane should not exceed 1.5°. Horizontal aperture angle should not exceed 10° within the radius of 200–300 m. Within a radius of 300 m, it should not exceed the height of the railway, and the obstacles relative to the vertical plane of the opening angle should not exceed 2.5°. As a reference point to DVOR antenna base center and as a reference plane to the antenna reflector net plane, the radius of 200 m height datum should not exceed 35 kV (or more) of high-voltage transmission lines, and the radius of 500 m of the plane should not exceed the height of 110 kV (or more) of high-voltage transmission lines. The DVOR signal interference protection ratio is 17 dB within the coverage area of FM radio, and the industrial, scientific, and medical equipment interference protection ratio is 14 dB, and other active interference protection ratio was 20 dB.

7.4.2.7 DME Station Requirements

The operating band of DME station is 962–1213 MHz, which provides continuous distance information to aircraft equipment and guides aircraft along the selected route (line) to flight arrival and departure and approach. Minimum signal field strength within the coverage area of conventional DME is 689 µV/m, and the minimum signal strength outside 13 km from DME station within the coverage area of precision DME is 689 µV/m, and the lowest signal strength within 13 km from DME station within the coverage area of precision DME is 3453 µV/m, in the lowest signal field strength at the approach reference point is 6140 µV/m.

Within the signal coverage area of DME, the protection to a variety of active interference is 8 dB. If DME is fitted with LOC, GS, or DVOR together, the same EME requirements will be claimed with the corresponding device. If DME is fitted as a separate desk, the center point of the antenna will be a reference point, and the center plane of the antenna will be a reference plane; there should be no obstacle within a radius of 50 m beyond the reference plane, there should be no obstacles beyond a radius of 50 m vertical plane opening angle of 3°, which should not have transmission lines of 110 kV and above of high voltage within 500 m.

7.4.2.8 Precision Approach Radar Station Requirements

Precision approach radar transmit horizontal and vertical scanning beams alternately and receive the reflected echoes of aircraft, mainly as a measuring device for guiding the aircraft approach and landing. It is usually arranged in flat areas or outside the airport landing road; the distance from the runway centerline is 120–225 m, retreat away from the landing point is not less than 915 m, and the angle of the connection point of the landing runway centerline is formed less than 9°. Operating frequency is 9370 MHz ± 30 MHz.

It should be flat and open around radar stations, and within the coverage area, there should be no obstacle in less than 500 m from the antenna as a benchmark higher than 0.5° opening angle.

7.4.3 Study Case

A straight line from the radio test site to an airport is 25 km, and the highest point between the site and the airport is intermediate 514 m above sea level. The main operating parameters of the transmitting device are as follows:

Operating Frequency	Maximum Transmit Power (W)	Antenna Form	Gain (dBi)	Work Type
2–30 MHz	125	Omnidirectional	3	Peer to peer
30–512 MHz	50	Omnidirectional	3	Hopping networking
610–960 MHz	3	Directional	14	Peer to peer
1350–1850 MHz	3 W	Directional	17	Peer to peer
4.4–5 GHz	20	Directional	21	Peer to peer
5.625–5.825 GHz	2	Directional	21	Peer to peer

These bands are able to work for the frequency range of the device and do not represent the frequency of tests used to communicate, and which is crossed with air navigation frequencies in the VHF range. The communication frequency of the radio test site is irrelevant to aeronautical navigation and communication frequencies, so there are no co-channel interference problems, according to the formula

$$E = \left(\frac{444 \times 10^3 \sqrt{P_t \cdot G_t}}{d} \right) \times \sin\left(\frac{2\pi \cdot h_t \cdot h_r}{\lambda \cdot d \times 10^3} \right)$$

where
  E means the space wave field strength
  P_t means the transmitter power
  G_t means antenna gain
  d means the distance between receive antenna and transmit antenna
  h_t means the height of the transmit antenna
  h_r means the height of the receive antenna
  λ means the radio wavelengths ($\lambda = c/f$)
  f means the RF

Protection rate of field strength can be calculated as follows:

Device Type	Frequency Range (MHz)	Aeronautical Communications Lowest Signal Strength (dBµv/m)	Interfering Signal Strength (dBµv/m)	Aeronautical Communication Requirements of Protection Rates (dB)	Actual Protection Rates (dB)
LOC	108–111.975	32	2.6	20	29.4
DVOR	108–117.975	39	2.6	20	36.4
GS	328.6–335.4	52	12	20	40
MB	75	63	−0.7	23	63.7

Interference protection distance can be calculated from the formula

$$d = 30 \times 10^{\left(\frac{E_{30} - E_S + R}{20A} \right)}$$

where
  d means the protection distance
  E_{30} means the equipment interference allowed values
  E_S means the signal strength of protective equipment
  R means the protection rate
  A means the attenuation coefficient

Interference protection distance can be calculated as follows:

Device Type	Frequency Range (MHz)	Device Interference Allowed Values dB (µv/m)	Minimum Distance from the Airport	Actual Linear Distance
LOC GS DVOR	108–400	40	1687 m (the lowest frequency)	25 km
			3308 m (the highest frequency)	

Based on the aforementioned analysis and calculation results, the radio test site for field emission device protection rate to airport navigation equipment is more than the required value 9 dB, and the protective distance is greater than the required value of 15–20 km. Thus, the radio test site does meet national standards and industry standards.

References

1. Walen, D. *Lightning and HIRF Protection*. Washington, DC: Federal Aviation Administration, 1998.
2. National Standardization Management Committee, Electromagnetic environment require- ments for aeronautical radio navigation stations, GB6364-2013, China Standardization Press, 2013.
3. Walen, D. *Aircraft Electromagnetic Compatibility*. Washington, DC: Federal Aviation Administration, 2002.
4. Ely, J. J. Electromagnetic interference to flight navigation and communication systems: New strategies in the age of wireless. Reston, VA: American Institute of Aeronautics and Astronautics, 2005.
5. NTIS. High-intensity radiated fields (HIRF) risk analysis. Springfield, VA: National Technical Information Service (NTIS).

8

Analysis and Modeling of the QoS Mechanism in ATN

Douzhe Li and Zhao Li

CONTENTS

8.1 Introduction

From the overall point of view, the aeronautical telecommunications network (ATN) is a mobile cellar wireless communication system that aims at transmitting both real-time and non-real-time data. Quality of service (QoS) is an important technique that can ensure the transportation of traffic with special requirements. In the ATN/IPS environment, QoS is a mandatory requirement that should be deployed since the wireless bandwidth is limited and is shared by multiusers and many types of services. QoS comprises requirements on all the aspects of a connection, such as service response time, loss, signal-to-noise ratio, crosstalk, echo, interrupts, frequency response, loudness levels, and so on.

For a thorough review of QoS and discover its application in ATN, first, in Section 8.2, we introduce QoS generally and then elaborate on the per hop behaviors (PHB) and class definition in ATN/IPS. Considering the recently developed L-band data link technology (i.e., L-DACS1), the QoS support in link layer will be introduced in Section 8.3. Some QoS protocols that are used in the network layer will be discussed in Section 8.4.

Routing is regarded as one of the key technologies for realizing QoS. By means of math-ematical theory, QoS routing is often modeled as a nondeterministic polynomial time (NP) complete problem and often solved by heuristic methods. In Section 8.5, some particular mathematical model that could be used to achieve the QoS requirements is discussed. Except well know Unicast problem, considering multicast service such as weather fore-casting and aeronautical chart updating will be widely used in ATN, we will discuss some particular multicast routing algorithms in Section 8.6.

8.2 Overview

Network resources (bandwidth, error rate, data transmission speed, etc.) are limited in any type of communication network, but at the same time, the intrinsic characteristic of a network is sharing the resource and transport data of multitude of applications, includ-ing high-quality graphics and delay-sensitive data such as real-time voice. In an ATN/IPS environment, various types of services (include legacy services that are defined in Doc 9705[1] and newly defined VOIP service) will simultaneously transmit data by utilizing ATN.

How to utilize the resources effectively and sufficiently is an important problem in designing and managing a particular network. To solve this problem, we introduce the definition of QoS. QoS is a set of techniques to analyze and manage network resources and is the metric of overall network performance, particularly the performance seen by the users in the network.

Bandwidth-intensive applications stretch network capabilities and resources, but also complement, add value, and enhance every business process. Networks must pro-vide secure, predictable, measurable, and sometimes guaranteed services. Achieving the required QoS by managing the delay, delay variation (jitter), bandwidth, error rates, throughput, and packet-loss parameters on a network becomes the key point to a success-ful end-to-end business solution.

As we have seen from Chapter 2, there are four layers in TCP/IP protocol stack infra-structure. We only consider two layers, one of them is the IP layer, also called Layer 3 (L3); the other is link layer (we only consider media access control [MAC] sublayer in link layer), also called Layer 2 (L2). Data transportation between L2 and L3 is dependent on an inter-face, for example, in L-DACS1 protocol, the interference is named SNDCP (subnetwork dependent convergence protocol, the network layer adaptation service of the SNDCP shall provide functions to transfer network protocol data units (N-PDUs) transparently over L-DACS1 subnetwork). In the ATN environment, L2 and L3 all define QoS requirements (L2-QoS and L3-QoS), but there are some differences between them. The main difference is, L2-QoS depends on the specific data link protocol and L3-QoS is almost a common standard and self-contained in IPv6. The detail will be discussed in Section 8.4.

In order to facilitate later explanation, first we will explain the related QoS mechanism in the link layer (L2-QoS). L-DACS1 is one of the promising candidate data links for future terrestrial aeronautical communication; understanding the QoS in L-DACS1 can help us to be more familiar with the L2/L3 joint optimization methods that have appeared in many papers.

8.3 QoS in Link Layer (Layer 2)

8.3.1 A Brief Introduction of L-DACS1 Data Link Protocol

Action Plan 17 (AP17)[2] is an internationally sponsored effort to investigate various possible candidate technologies which can be used to support future aeronautical communications. Some of the considered and evaluated technologies operate in the L-band, supporting the [COCRv2][3] requirements. But none of them can fulfill each requirement perfectly. The main reason is operational spectrum compatibility, since an existing system in the L-band will cause interference if there is overlap with another system. However, the analysis of an existing system and desired features of a new system can help us to develop something suitable or design a technology to mitigate the interference.

Given these features and the most promising candidates, two technical solutions of L-band digital aeronautical communications system (LDACS) were developed. Before the final choice of a suitable data link from these two options, some further evaluation is required.

One choice, L-DACS1, is a frequency division duplex (FDD) configuration based on orthogonal frequency division multiplexing (OFDM) technology, it has some advantages, such as high data rate (up to 2.6Mbit/s) and a more robust multipath channel environment. Also it has reservation based access control, and can support upper layer advanced network protocols. L-DACS1 is evolved from broadband aeronautical communications (B-AMC) and TIA-902 (P34) technology and has a close relationship with them.

Another choice is L-DACS2, a time division duplex (TDD) based system, which is derived from all-purpose multicarrier aviation communication system (AMACS) technologies. The modulation of L-DACS2 is binary modulation (e.g, continuous-phase frequency-shift keying, CPFSK). This type of modulation exists in some commercial systems such as GSM or universal access transceiver (UAT) system. L-DACS2 can also support QoS requirements and upper layer network adaptation. Its bandwidth is narrower than L-DACS1, so the data rate is lower.

AP17 and SESAR proposed that some activities should be taken before 2020, the deadline for a final decision, to further consider the proposed two LDACS options:

- Develop detailed specifications for L-DACS1 and L-DACS2
- Develop and test L-DACS1 and L-DACS2 prototypes
- Assess the overall performance of L-DACS1 and L-DACS2 systems

Specific EUROCONTROL contracts covered the initial activities to develop detailed specifications for the L-DACS1 (Project ID 15.02.04.) and L-DACS2 system.

The interference of L-DACS is mainly from distance measures equipment (DME), but also there is other interference such as from Joint Tactical Information Distribution System (JTIDS) which is based on frequency hopping technology. AP17 suggests that spectrum compatibility investigations should be conducted in a consistent way, i.e., the testing environment must be standardized to give a fair conclusion.

Note: Another EUROCONTROL contract has focused on the development of the interference scenarios to be investigated and the definition of acceptability criteria for each scenario.

8.3.2 QoS and RRM (Radio Resource Management) in L-DACS1

The L-DACS1 data link protocol has several inner modules (i.e., functional blocks).[4,5] These modules are connected by some messages whose transmission action is indicated by a black double arrow "↔", as we can see from Figure 8.1. In the actual system, the message is realized by a PDU (physical data unit), which is a block of bytes (octet) that each bit has specific meaning. Different modules have different PDU, as described in the following.

PHY: Physical layer based on OFDM[16] that realizes the physical frame construction and deconstruction, coding, decoding, modulation, and demodulation. The message that PHY layer uses is PHY PDU. PHY is out of the scope of this chapter.

MAC: Media access control layer provides an abstraction logical channel to hide the PHY layer; the message that MAC layer uses is MAC PDU.

DLS: Data link service layer. It may be utilized by the link management entity (LME) for the conveyance of signaling/management data and the SNDCP for the conveyance of SNDCP data PDUs or signaling. The message that DLS layer uses is DLS PDU.

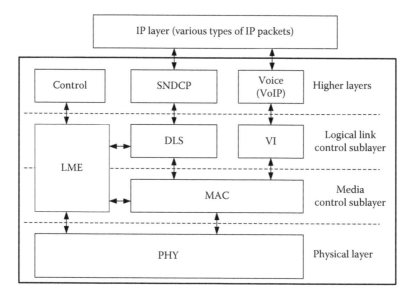

FIGURE 8.1
L-DACS1 protocol stack and its upper IP layer.

SNDCP: Network layer protocols are intended to operate over services provided by a variety of subnetworks and data links. L-DACS1 subnetwork can support several network layer protocols while providing protocol transparency for the user of the subnetwork service. The message that SNDCP uses is SN-PDU.

LME: Not having a PDU format definition, LME can be seen as the manager of L-DACS1. LME controls PHY through interface, and also conducts configuration, resource management, and mobility management.

We have to keep in mind that all aircraft stations in the same cell are controlled by a single ground station. Before an aircraft station sends data, it should first request resources (the request message is included in the multiframe of physical layer) from its controlled ground station and the ground station will give a time slot in the multiframe (for the definition of multiframe, please refer to L-DACS1 specification). The resource allocation is contention free; each aircraft station will have its own static slot in each transmission.

The Data Link Layer (DLL) provides QoS assurance according to COCRv2 requirements. Multiplexing of different service classes is possible. Except for the initial aircraft cell-entry, medium access is deterministic, with predictable performance. Optional support for adaptive coding and modulation is provided as well.

The DLS offers its services prioritized, that is, in different classes of service. Service classes map directly to priorities. The service class of a request shall be used by the ground station to determine the order and size of resource allocations. Within the data link service the service class is used to determine the precedence of concurrent service requests. The classes of service supported by DLS are displayed in Table 8.1.

Service class DLS_CoS_7 shall designate the service class with the highest priority and DLS_CoS_0 shall designate the service class with the lowest priority. DLS_CoS_7 shall be reserved for LME signaling. DLS_CoS_6 shall be reserved for the packet mode voice service (VoIP).

QoS mapping between applications and link technology service classes is one of the considerations for the scheduling mechanism. On the one side, L-DACS1 supports 8 different service classes, and on the other side, 12 different application level classes of service are defined.[3] The mapping between application class of service (CoS) and L-DACS1 service classes are not studied here, and we only assume two different application CoSs are mapped to two different L-DACS1 service classes.

TABLE 8.1

DLS CoS Mapping onto ATN/IPS Application Levels

Class of Service	Priority	Comment	
DLS_CoS_7	Highest	Reserved for LME	Mapping to LME control signals and VoIP
DLS_CoS_6		Reserved for packet mode voice	
DLS_CoS_5			Mapping to other applications according to COCRv2
DLS_CoS_4			
DLS_CoS_3			
DLS_CoS_2			
DLS_CoS_1			
DLS_CoS_0	Lowest		

8.3.3 Algorithms for Radio Resource Management (RRM)

According to L-DACS1 specification, both FL (forward link, means the data transmission from ground station to aircraft station) and RL (reverse link, means the data transmission from aircraft station to ground station) resources shall be assigned by the GS's (ground station's) RRM function. Implementation specific details of the resource allocation procedures are not involved in L-DACS1 specification, so several popular RRM algorithms will be discussed in this section.

For better understanding, the readers should be familiar with the steps of L-DACS1 radio resource request and allocation.

If several aircraft belonged to a ground station, the resource request is periodically sent by every aircraft station to indicate whether it has data to be transmitted. Ground station will periodically receive the requests from aircrafts that are under its control, then the ground station should calculate and allocate available resources (i.e., bandwidth) to each user (i.e., aircraft). In the RL, L-DACS1 only supports one type of bandwidth request mechanism: explicit contention-free bandwidth request and response. An aircraft will have its privileged time slot for data transmission; if the ground station does not allocate a specific slot for an aircraft, it cannot transmit data in that period. In this mechanism, each aircraft explicitly sends its bandwidth request in the dedicated control channels (DCCH) slot in order to receive resources for the RL transmission to the BS in the upcoming data slot (i.e., RL Data).

The IP datagram is first packaged through SNDCP interface to force an SN-PDU. Then, DLS will separate SN-PDU into several queues according to different CoS. Finally, each queue will be segmented into several segments to force DLS_PDU. A critical point is the size of segmentation that is based on the resource allocation that is described as follows.

A resource request is transmitted by a MAC PDU that can be seen in Table 8.2; there are two types of requests, *single resource requests (SRSC_RQST)* and *multiple resource requests (MRSC_RQST)*. Each type of resource request should indicate the CoS type and how many PHY PDUs of this CoS is requested. The resource request MAC PDU will be converted to a PHY PDU and sent by a logical channel called DCCH; actually, the logical channel is sited in a physical time slot that is called multiframe (MF). The details of MAC PDU can be found in Chapter 8 of L-DACS1's specification.[4]

After the ground station has received every user's resource request, it will use some algorithms to allocate resources. The allocation result will be sent to aircraft by a MAC

TABLE 8.2

Resource Request Format

Field	Size (bit)	Description
D_TYP	4	Resource request
SC	3	Service class
NRPP	8	Number of RL PHY PDUs
D_TYP	4	Resource request
ENT	3	Number of entries (the number of CoS)
SC_1	3	Service class
$NRPP_1$	8	Number of RL PHY PDUs
...	...	...
SC_8	3	Service class
$NRPP_8$	8	Number of RL PHY PDUs

TABLE 8.3

Resource Allocation Format

Field	Size (Bit)	Description
C_TYP	4	RL allocation
SAC	12	Subscriber access code
RPPO	8	RL PHY-PDU offset
NRPP	8	Number of RL PHY PDUs
CMS	3	Coding and modulation scheme
RES	9	Reserved
CRC	4	Cyclic redundancy checksum

PDU, which can be seen in Table 8.3. It indicates the ID of a specific aircraft (subscriber access code) and the resources (how many PHY PDU can be sent to the aircraft in the next MF for both RL and FL).

It should be clear that there are two features in L-DACS1 resource allocation:

1. Data queues (aircraft) and scheduler (ground station) are located in separate compartments; this is different from other wireless communication data link protocols.
2. Normally in every MF, the DLS entity where the queues are located sends bandwidth requests for every user to the LME entity where the scheduler is running.

While a strict priority scheduling mechanism or weighted fair queuing can be applied among different service classes at the first level according to Table 8.1, another scheduling algorithm is used for each service class at the second level. Three different scheduling mechanisms are investigated for L-DACS1 by Serkan Ayaz[6]:

1. Modified deficit round robin with fragmentation (MDRR)
2. Fair-share scheduling (FSS)
3. Randomized user selection scheduling (RANDUS)

Based on the above mentioned schemes, Figure 8.2 shows the main L-DACS1 scheduling architecture for the RL and FL in L-DACS1. The scheduling algorithm for FL and RL is running in the LME block and packets are queued in the DLS block.

Strict priority scheduling or weighted fair queuing mechanisms can be used at the first stage depending on the usage of application types. If currently defined ATN/IPS applications are only used in the network, then strict priority scheduling is enough since the high priority message volume is significantly smaller compared to the low priority message volume. In this case, we do not expect any bandwidth starvation problem for low priority messages. However, if VoIP with high priority is accepted to be used in the future ATM then it might cause bandwidth starvation. In this case, usage of a weighted fair queueing scheduling mechanism is more acceptable since it allocates a certain portion to all service classes from available bandwidth depending on its priority level.

For this reason, deficit round robin (DRR) is modified in a way that it operates on the bandwidth request of each user (sent by DLS) and not based on the size of the head-of-line packet. In the implementation, the quantum size (Qsize) is selected equal to the packet size. The modified algorithm is able to run for both links (i.e., FL and RL) separately.

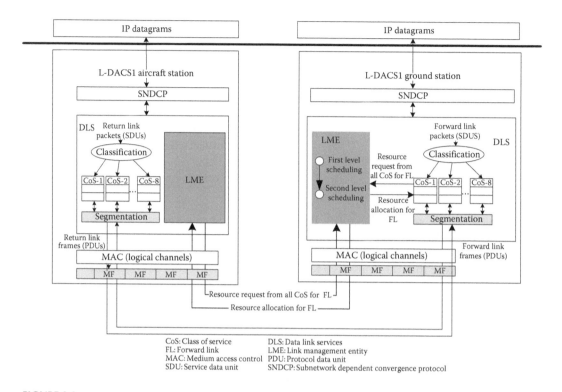

FIGURE 8.2

L-DACS1 link scheduling architecture. (Adapted from Ayaz, S. Mobility and radio resource management in future aeronautical mobile networks. PhD thesis. Erlangen, Germany: University Erlangen-Nürnberg, 2012)

The author not only describes the algorithms, but also given some performance evaluation. The result is, according to experiments, the tradeoff can be shown between fragmented packet transmissions (which is directly proportional to the overhead) and fairness among different users. On the one side, FSS provided best fairness results while causing the highest number of fragmented packet transmissions, on the other side RANDUS provided worst fairness results with the least fragmented packet transmissions. From this perspective, DRR provided a good balance between these two criteria.

The resource allocation problem can be described as the total bandwidth W that will be allocated to N users. Ground station knows each user's situation (how many CoS and how many PHY PDUs for each CoS). FSS and RANDUS are relatively simple, so we will first describe them.

8.3.3.1 Fair-Share Scheduling

FSS is designed to allocate resources equally to all users. In our case, when the LME receives resource requests of all active users, it directly divides the total available bandwidth by the number of users and assigns the corresponding portion to each user. There are two important considerations for the algorithm:

1. In case the resource request of a user is less than the available bandwidth for a single user then the remaining resources are uniformly distributed among the other resource requesting users.

2. Similarly, in case there are some remaining bytes or PDUs left due to division of total available resources to the number of resource requesting users, then these remaining bytes or PDUs are also uniformly distributed among the resource requesting users.

8.3.3.2 Randomized User Selection Scheduling

The idea of RANDUS algorithm is to assign resources to users randomly considering uniform distribution in every MF as shown in Algorithm 6 of Serkan Ayaz's PhD thesis[6]. It is a basic scheduler that does not consider any fairness criteria when distributing resources among different users. In our implementation, random shuffle function of C++ STL algorithms library is used for randomizing the resource requesting users.

8.3.3.3 Modified Deficit Round Robin with Fragmentation

DRR, also called deficit weighted round robin (DWRR), is a scheduling algorithm for the network scheduler. DRR is a modified weighted round robin and was proposed by M. Shreedhar and G. Varghes in 1995. There are two important concepts in DRR we should know:

- *Deficit counter (DC)*: The number of bytes that a flow is allowed to transmit when it is its turn
- *Quantum size (Qsize)*: The number of bytes that is added to the deficit counter of flow in each round (quantum = amount of credit per round)

Applying different quantum to different flows, weighted fairness can be achieved. Selecting the size of DC and Qsize is dependent on the specific application.

When a packet arrives to the queues, its flow number is checked first. The flow number can be seen from the Netscape server application programming interface (NSAPI) field in SN PDU. In case the flow number is in the list (i.e., active list), the packet is just enqueued corresponding to the specified flow number. In case the flow number is not in the list, it is inserted in the active list and the packet is enqueued corresponding to the specified flow number. In our case, when the LME module receives resource requests of all users, it performs similar operation for the insertion of the user to active list and setting deficit counter (DC) to zero.

Afterwards, the dequeuing operation of packets takes place while the algorithm is running according to the original algorithm[7] such that it first checks whether the DC is greater or equal to the size of the head-of-line packet. If this is the case, the algorithm performs the dequeuing operation, sends the packet, and decreases the DC by packet size. In the other case, the algorithm just increases the DC by a quantum size (Qsize) and does not send the packet. Since the algorithm needs to know the size of the head-of-line packet, it requires certain modification for L-DACS1. The reason is that, in L-DACS1, queues and scheduler are located in separate compartments.

In L-DACS1, normally in every MF, the DLS entity where the queues are located sends bandwidth requests for every user to the LME entity where the scheduler is running. For this reason, we had modified the DRR in a way that it operates on the bandwidth request of each user (sent by DLS) and not based on the size of the head-of-line packet as shown in Algorithm 4 of Serkan Ayaz's PhD thesis[6] (i.e., lines 11–51). In our implementation, we select the Qsize equal to the packet size. The modified algorithm is able to run for both links (i.e., FL and RL) separately.

It also needs to be mentioned that the Qsize is selected in terms of bytes on the FL and in terms of PDUs on the RL according to the bandwidth requests from DLS in BS and DLS in aircraft, respectively. For real-time services (e.g., VoIP), the quantum size is selected as 103 B on the FL and 8 PDUs (which makes 112 B for the lowest coding modulation scheme) on the RL. For file transfer on the FL, we select a quantum size of 1091 B. Since the resource requests of different CoSs are sent separately, it is possible to differentiate which quantum size is used for which CoS traffic.

8.3.3.4 RRM Time Complexity Analysis Methods

Time complexity analysis for three methods is analyzed in Serkan Ayaz's PhD thesis.[6] It can be found that the time complexity of the scheduling process is correlated with two components but is independent from the scheduling algorithm. One of the components, LME, will receive the resource request messages that are transmitted by DLS. Then resource schedule algorithms will be used to process different requests. Assume there are N users in the system that are requesting resources. The algorithm needs to process each user's request, therefore the complexity of messages is $O(N)$.[6]

Another component is the dequeuing operation. After the resources have been assigned to each user by the algorithm, if the results indicate that M ($M \leq N$) users have been entitled to transmit data, then the data packages corresponding to M users will be dequeued. Each user's data packages in DLS should be searched before they can be dequeued. In a practical system, we can use an algorithm library to realize the operation. Here, in order to measure the complexity, we take the "find" function of a C++ STL as an example. The "find" function can be utilized to search data packages for each user, as we know the complexity of the "find" function is $O(\log N)$. If $M < N$, the total complexity is $O(M) \times O(\log N)$, the maximum complexity is $O(N \log N)$, if $M = N$. In FL, resource assignment is by byte, while in RL, resource assignment is through PDU package.

The performance of different scheduling algorithms is based on considering two different ATM applications. We restrict our work to scheduling algorithms running at the second level and focus on two different service classes separately. In the first class of service, we consider file transfer of graphical weather information (i.e., WXGRAPH service). In the second class, we consider VoIP as an example for real-time service. Using the integrated L-DACS1 model, we thoroughly analyzed the performance of three different scheduling algorithms for each service class in terms of delay and bandwidth fairness.

8.4 QoS in Layer 3: DiffServ per Hop Behaviors and Class Definition

8.4.1 QoS Support in IPv6

Several features were added to the IPv6 specification in addition to 128-bit addressing as the IPv6 specification made its way through the IETF committee process. The added features include levels of assured service, enhance security, and improved reliability.

QoS is an important term and an emerging feature of modern networks. IPv4 networks typically give each and every packet a *best level of effort* service, even if the content of every packet is not really important or time-sensitive.

An IPv4-based system has no way to differentiate between data payloads that are time sensitive, such as streaming video or audio, and those that are not time-sensitive, such as status reports and file transfer. Streaming audio and video application are very sensitive to a delay of a few packets—lips move without sound or picture break up—but IPv4 has no way to prevent those problems.

If a packet is lost in transit, TCP recognizes the loss and requests a retransmission, but only after an inevitable delay. The single delayed TCP packet is probably part of a much larger packet of audio or video data, so the entire big packet is delayed and probably thrown out because the smallest part did not arrive on time.

IPv6 provides a way for applications to request handling without delay throughout the wide area network (WAN). The term often used to describe this is low latency. Streaming audio and video requires low latency through high priority. To prevent a breakdown in the scheme, various applications can share a connection via priority level.

- Level 0—No specify priority
- Level 1—Background traffic (news)
- Level 2—Unattended data transfer (email)
- Level 3—Reserved
- Level 4—Attended bulk transfer (FTP)
- Level 5—Reserved
- Level 6—Interactive traffic (Telnet, Windowing)
- Level 7—Control traffic (routing, network management)

Packet fragmentation is a major source of packet delay, or high latency, under IPv4. Each device attached to a network has a payload data limit set inside the Ethernet packet. If a program is generating streaming data, such as video or audio, the data stream will be split up into a string of packets, each carrying the maximum payload.

With different devices, these payload sizes are set differently and it is possible that between the originating source and the destination, some transmission path, particularly an asynchronous transfer mode (ATM) link, will have a smaller payload size.

The ATM equipment will chop the already fragmented stream of data into even smaller pieces. An ATM switch could divide the data carried in one single Ethernet packet into 20 ATM cells. Somewhere in the dividing and rebuilding of all the data, it is likely that a cell or a packet will be dropped or delayed.

IPv6 uses a more sophisticated approach to handle data from programs requesting priority handling. The originating device will query the destination in order to determine the maximum size of the payload that can be handled across the complete route. Then it adjusts its own parameters and will not load the originating packets with more data than the network can handle.

This approach reduces fragmentation and latency but can also result in inefficient utilization. The tradeoff is that with shorter payloads, it will achieve a higher bandwidth with prompt arrival.

QoS functionality will have to be included on every networked device in order to be implemented. Unavailability of functionality on certain devices will cause it to fallback to a standard handling with just an additional layer to pass through.

Currently IPv6 provides support for QoS marking via a field in the IPv6 header. Similar to the type of service (ToS) field in the IPv4 header, the traffic class field (8 bits) is

TABLE 8.4

IPv6 Packet Header

Version = 6	Traffic Class	Flow Label
Payload length	Next header	Hop limit
Source address		
Destination address		

available for use by originating nodes and/or forwarding routers to identify and distinguish between different classes or priorities of IPv6 packets (Table 8.4).

The *traffic class* field may be used to set specific precedence of differentiated services code point (DSCP) values. These values are used in the exact same way as in IPv4.

The key advantage with the *flow label* is that the transit routers do not have to open the inner packet to identity the flow, which aids with the identification of the flow when using encryption and other scenarios. In the ATN environment, IPS nodes shall set the flow label field of the IPv6 header to zero, as it is not used in the ATN/IPS.

IPv6 also has a 20-bit field known as the *flow label* field (RFC 3697). The *flow label* enables per-flow processing for differentiation at the IP layer. It can be used for special sender requests and is set by the source node. The flow label must not be modified by an intermediate node.

L3-QoS uses ToS (type of service) and deals with IP address like

1. Classification/prioritization of packets in forwarding path based on DSCP IP header field
2. Policy and allocation of priorities along the transmission path

Each router supporting DSCP needs to be configured accordingly (priorities).

The routers along a path (source to destination) do not store the state about the flow (e.g., number of packets already transmitted). Instead the routers apply a policy on each packet individually (Figure 8.3).

- *Classifier*: Classifies packet into an internal class. For intermediate routers, the classifier may be missing (packet already classified by AS ingress router)

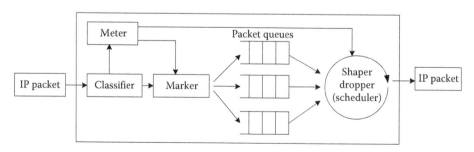

FIGURE 8.3
General logical architecture of a router.

- *Marker*: Measures the temporal properties of the packet stream selected by the classifier and instructs the marker and shaper/scheduler to treat a specific packet accordingly (e.g., drop a packet that is out of profile, i.e., the stream already used 100% of the available bandwidth). Marks packets according to the class defined by the classifier.
- *Shaper/scheduler*: Extracts packets from queues according to a local policy and sends the packet.

ToS field was too inflexible and redesigned to a single field named DSCP.

DSCP contains a number that indicates the PHB to be applied to the IP packet (Figure 8.4).

Packets are classified (and DSCP field marked) at the ingress into a domain (e.g., autonomous system or administrator domain in ATN/IPS).

Intermediate routers in domain B prioritize packets according to the DSCP field in IP header.

Domain B egress router shapes and schedules packets. The process can be seen in Figure 8.5.

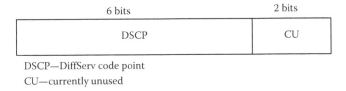

DSCP—DiffServ code point
CU—currently unused

FIGURE 8.4
New definition for IPv6 QoS label in RFC2474.

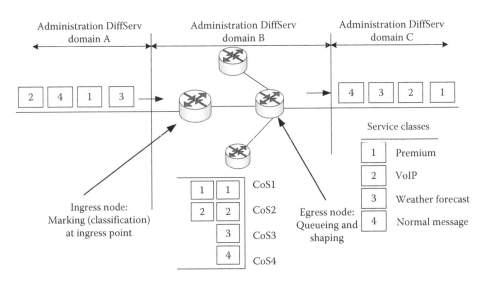

FIGURE 8.5
Illustration of DiffServ domain effect.

8.4.2 QoS in ATN/IPS Context

ATN/IPS communication service providers are likely to make use of the same IPS infrastructure for ATN and other non-ATN defined applications; for example, Air Traffic Service Message Handling System (ATSMHS) and surveillance data. Sharing of resources can be at different levels, ATN/IPS applications can use the same type of CoS as non-ATN applications over the same IP-routed infrastructure. Alternatively, ATN/IPS communication service providers may only wish to share the same physical infrastructure and operate a VPN per service; in this case a separate CoS model can be applied to each virtual private network (VPN) service, one being the ATN/IPS. Fundamentally, ATN/IPS communication service providers have flexibility in how they enable CoS for the ATN/IPS over their infrastructure.

For CoS definitions, it is essential that ATN/IPS traffic is sufficiently qualified in order to properly mark ingress traffic. As the IP packet enters the network core, PHBs are enforced, depending on the packet marking. ATN/IPS communication service providers will need to handle unmarked or premarked ingress traffic and be prepared to mark or remark the traffic before it is routed over their infrastructure. The internal techniques, mechanisms, and policies to enforce the PHB within the communications service provider networks are considered out of the scope of the ATN/IPS.

8.4.3 ATN/IPS PHBs/CoS

The ATN/IPS is to support legacy ATN applications over the IPS. Currently, this intended support covers CM(DLIC), FIS(ATIS), CPDLC, ADS-C, and ATSMHS. Indeed, DIR is only specified for ATN/OSI and it is foreseen that AIDC will be implemented through regional solutions. As each ATN application is mapped to a given CoS, the dynamic support of different priorities per user message category is not considered.

Table 8.5 provides an example of an administrative domain that supports several applications and CoSs labelled very high, high, normal, and best effort.

In order to mark ingress traffic, the ATN/IPS provider has several means to identify the traffic: destination transport port number, IP source address, and IP destination address.

ATN/IPS is not recommended to use the DSCP/ToS value set by the application or prior communication service provider, because the DSCP/ToS field may be incorrectly configured or unknown, and will not be the optimum approach.

8.4.4 DiffServ Code Point Values

PHB is indicated by encoding a 6-bit value—called the DSCP—into the 8-bit differentiated services (DS) field of the IP packet header. The DSCP value of the field is treated as a table index to select a particular packet-handling mechanism. This mapping must be configurable and administrative domains may choose different values when mapping code points to PHBs. However, it is widely accepted that DSCP value 101110 refers to EF (expedited forwarding).

Table 8.6 provides an example of mapping DSCP values to ATN/IPS PHBs where a number of applications share the same IP network infrastructure. In this table, air–ground applications have been assigned with the special class selector code points as specified in Document 9880 for the ATN IP SNDCF, but within the ATN/IPS it would be better to make use of AF PHBs to avoid any interaction with legacy applications that make use of IP precedence.

TABLE 8.5

ATN/IPS Priority Mapping to Classes

Priority/Application Mapping			Traffic Identification (Ingress)		
Class (CoS Type)	Drop Precedence	ATN Priority	ATN Application	TCP/UDP Port	IP Address
Very high (EF)			Voice (VoIP)	RTP numbers 16384-32767	—
High (AF)	1	0	—	—	—
		1	—	—	—
		2	—	—	—
		3	ADS-C	TCP 5913 UDP 5913	The source or destination address will be part of a reserved address space assigned to mobile service providers
			CPDLC	TCP 5911 UDP 5911	
Normal (AF)	1	4	AIDC	TCP 8500[a]	
			FIS(ATIS)	TCP 5912 UDP 5912	The source or destination address will be part of a reserved address space assigned to mobile service providers
	2	5	METAR	—	—
	3	6	CM(DLIC)	TCP 5910 UDP 5910	The source or destination address will be part of a reserved address space assigned to mobile service providers
			ATSMHS	TCP 102	
		7			
Best effort (default)		8–14	—	—	

Source: ICAO, *Manual for the ATN Using IPS Standards and Protocols*, Doc 9896, Draft Version 21, 2013.

[a] This is applicable when OLDI/FMTP is used as a means to enable AIDC services.

8.4.5 Traffic Characterization

Traffic characterization is a means to express the type of traffic patterns, integrity, and delay requirements. It provides further information to the communication service provider in order to fully meet the user requirements within a specific network operation. Typically, traffic characterization information is part of the contracted service level agreement in which further parameters are defined such as service delivery points, service resilience, required bursting in excess of committed bandwidth, service metric points, mean time to restore (MTTR), and port speeds.

Table 8.7 provides an example of traffic characterization for ground–ground services, which is derived from the Pan-European Network Services (PENS) specifications.

8.4.6 QoS Support in Wired to Wireless MIPv6/NEMO

The above sections only consider the QoS problem in L2 and L3 separately; it is obvious that ATN/IPS is a complex and integrated mobile system and handover between two

TABLE 8.6

Example of DSCP to PHB Mapping

DSCP Value	PHB	Application
000000	CS0	Best effort
001000	CS1	
001010	AF11	AIDC
001100	AF12	
001110	AF13	
010000	CS2	CM
010010	AF21	ATSMHS
010100	AF22	
010110	AF23	
011000	CS3	FIS
011010	AF31	Voice recording
011100	AF32	
011110	AF33	
100000	CS4	CPDLC, ADS-C
100010	AF41	Voice signaling
100100	AF42	
100110	AF43	
101000	CS5	
101110	EF	Voice
110000	NC1/CS6	
111000	NC2/CS7	

TABLE 8.7

Example of Traffic Characterization

ATN Application	Average Message Length	Expressed Integrity	Jitter	Typical Bandwidth (Point to Point) (kbps)	Network Delay (1-way)
Voice (VoIP using G.729)	70 bytes	—	<15 ms	12	<100 ms
OLDI/FMTP (regional AIDC)	150 bytes	1 user corrupt message in 2000	N/A	10	<1 s
ATSMHS	3 kb	10^{-6} (in terms of 1000 bytes message blocks)	N/A	20	<5 s

ground stations is a common problem. How to keep the QoS (especially required by time-sensitive applications) during handover is attractive and challenging. In Chapter 3, we have introduced MIPv6 and four types of extensions such as PMIPv6, HMIPv6, FMIPv6, and NEMO. The purpose of these extensions is reducing the signaling time to expedite the data rate.

QoS performance in mobile environment depends on two aspects: The first is how the node mobility affects end-to-end QoS guarantees and the second is how to apply the existing QoS technologies in wired networks to wireless networks, that is, how to append mobility support to these solutions and how these solutions suit the wireless link characteristics.

As mentioned before, QoS is closely related to L2 and L3, so most of the study is to consider a hierarchical method (or cross-layer method, their meanings are almost the same) to solve this problem. We will provide some research achievement in this section to find a reasonable approach.

Lee et al. proposed a cross-layer hierarchical network mobility framework called Hi-NEMO for all-IP networks.[8] The advantage of Hi-NEMO is that it does not have triangular routing between an MNN and the CN. The design is resilient to error-prone transmission, and protocol supports fast QoS provisioning in the NEMO service domain. A QoS-incorporated handover (QoS-handover) is suitable to fulfill QoS requirements of real-time multimedia applications on a high-velocity vehicle and not just the reestablishment of network connectivity.

Kan et al.[9] proposed a new two-plane two-tier QoS architecture based on the advantages of IntServ and DiffServ. This chapter describes how the architecture guarantees an end-to-end QoS. Finally, the existing mobile IPv6 signaling such as binding update (BU), binding request (BR), and binding acknowledge (BA) is extended for QoS negotiation and advance resource reservation.

Ayaz gave the method to speed up the handover performance in an MIPv6 environment.[6]

Kong et al. have already proven by experiment that PMIPv6 is a promising candidate solution for realizing the next-generation all-IP mobile networks. As mentioned in Section 2.5.3, PMIPv6 is using LMA and MAG to construct an enhanced function to extend MIPv6. Liebsch et al.[10] proposed a new mobility option, the QoS option, for PMIPv6.

Although some of the studies do not focus on L-DACS1 and ATN/IPS environment, the concepts can be applied to ATN/IPS environment since ATN/IPS is also an all-IP network.

8.5 Basic Knowledge and Mathematical Model of QoS Routing

In RFC2386, QoS-based routing is defined as "a routing mechanism under which paths for flows are determined based on some knowledge of resource availability in the network as well as the QoS requirement of the flows." In short, it is a dynamic routing scheme with QoS consideration.

Before we introduce the QoS routing problem, two concepts that are correlated with ATN/IPS routing should be first described. For better understanding, an illustration is also shown in Figure 8.5

1. *Administrative domain*: From the view point of administrators, the ATN/IPS internetwork is composed of several interconnected administrative domains. An administrative domain can be an individual state, a group of States (e.g., an ICAO Region), an air communications service provider (ACSP), an air navigation service provider (ANSP), or any other organizational entity that manages ATN/IPS network resources and services (Figure 8.6).

2. *Autonomous systems (AS)*: From a routing perspective, interdomain routing protocols are used to exchange routing information between ASs, where an AS is a connected group of one or more IP address prefixes. The routing information exchanged includes IP address prefixes of differing lengths. For example, an IP

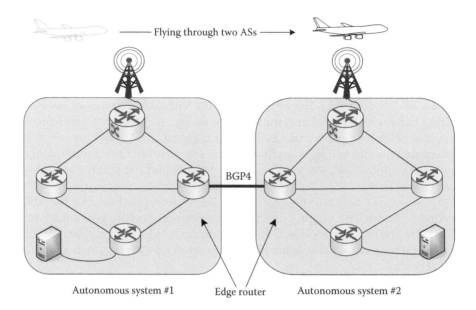

FIGURE 8.6
(See color insert.) ATN/IPS routing infrastructure.

address prefix exchanged between ICAO regions may have a shorter length than an IP address prefix exchanged between individual states within a particular region.

The ATN/IPS standard only specified the interdomain routing protocol (i.e., *border gateway protocol* (BGP-4) as specified in RFC 4271). The routing protocol within an AS is a local matter determined by the managing organization. Interdomain routing protocols can be used to exchange routing information among ASs. A single administrative domain may be responsible for the management of several ASs. BGP is a type of *best-effort* routing protocol; it will use the *shortest path* to the destination (shortest path does not necessarily mean the path with shortest physical distance. It may also mean the path with the least cost or fewest hop counts, for instance). In other words, they usually use single objective optimization algorithms that consider only one metric (bandwidth, hop count, and cost). Thus, all traffic is routed to the shortest path. Even if some alternate paths exist, they are not used as long as they are not the shortest ones. One drawback of this scheme is that it may lead to the congestion of some links, while some other links are not fully used.

ATN/IPS is expected to support a wide range of communication-intensive, real-time multimedia applications or non-real-time applications such as file transfer. The requirement for timely delivery of digitized audio–visual information raises new challenges. In order to ensure QoS performance, one of the key issues is solving QoS routing problem. It finds the best network routes with sufficient resources for the requested QoS parameters. The goal of routing solutions is twofold:

- Satisfying the QoS requirements for every admitted connection
- Achieving the global efficiency in resource utilization

There are two types of information transportation in ATN/IPS,[2] *unicast* and *multicast*. Unicast is a point-to-point transfer that sending messages to a single network destination identified by a unique address and *multicast* is group communication where information is addressed to a group of destination computers simultaneously. In the ATN, the need of sending the same information to multiple receivers is one of the main requirements of surveillance data distribution. Using a mathematical model can extract the essence and give a better solution.

8.5.1 Model of QoS Routing

A network can be modeled as a graph that is used to describe and calculate routing problems. A graph can be written as $G(V, E)$; nodes (V) of the graph represent switches, routers, and hosts. Edges (E) represent communication links. Since the bidirectional and asymmetric characteristic of most communication links, every link is represented by two directed edges in the opposite directions. The purpose of solving QoS routing can be described as

1. For unicast, finding a network path that meets the requirement of a connection between two end users

2. For multicast, finding a multicast tree rooted at a sender and the tree covers all receivers with every internal path from the sender to a receiver satisfying the requirement

Routing consists of two basic tasks. First, how to measure and collect network state information; second, how to compute routes based on the information collected. The performance of any routing algorithm directly depends on how well the first task is solved.

Figure 8.7 gives a brief visualization of the graph model in QoS routing; m and D is the information source and destination and a–d are nodes included in the network. The link state consists of residual bandwidth, delay, and cost. Each node should have a state. The node state can be either measured independently or combined into the state of the adjacent links. For the latter case, the residual bandwidth is minimally the link bandwidth and the CPU bandwidth; the delay of a link consists of the link propagation delay and the queueing delay at the node; the cost of a link is determined by the total resource consumption at the link and the node, such as charges of service providers.

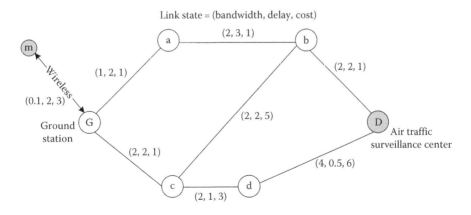

FIGURE 8.7
Abstract network state.

According to Figure 8.7, the unicast routing problem is defined as given a source node m, a destination node D, a set of QoS constraints C, and possibly an optimization goal, find the best feasible path from m to D, which satisfies C. The multicast routing problem is defined as given a source node m, a set of destination nodes T, a set of constraints C, and possibly an optimization goal, find the best feasible tree covering S and all nodes in T, which satisfies C. The two classes of routing problems are closely related. The multicast routing can be viewed as a generalization of the unicast routing in many cases. These two problem classes can be further partitioned into subclasses.

Mainly, there are three types of QoS-based routing algorithms, which are called hop-by-hop routing (also called distributed routing), source routing (also called path addressing) and hierarchical routing. "They are classified according to the way how the state information is maintained and how the search of feasible paths is carried out."[11]

- For hop-by-hop routing, the routing table list in each router only contains the address of the next device towards the destination. Therefore a router only forwards a data package according to the next-hop router. This method is the fundamental characteristic of the IP internet layer, and is utilized by the well known Routing Information Protocol (RIP) and most current Internet routing protocols.

- For source routing, the sender of package has partial or complete knowledge about the network, i.e., the global network state information can be obtained, so the source router can locally specify the path that the package will be used to reach the destination router. The advantage of this method is it is easy for troubleshooting and enhancing the trace-route. The disadvantage is the source does not have the capability to avoid congestion.

- Hierarchical routing is correspondence to flat routing (e.g. source routing) and is based on hierarchical addressing. That means the routers are arranged in a hierarchical manner, i.e., the actual routers are placed in the bottom level, and they can be grouped into several next layer nodes. Continuing this process recursively, we can build a large network. For example, most TCP/IP routing is a two-level hierarchical routing, i.e., each IP address consists of two parts, the network portion and the host portion. The gateway divides a package according to the network portion. An advantage of hierarchical routing is that it reduces the routing table size from n^2 to $\log n$, but the overhead is increased.

8.5.2 Routing Problem Model Description

ATN/IPS is a wireless network that needs QoS guarantees. However, the dynamic nature of ATN makes it more difficult to provide QoS, because it is hard to keep routing state information up to date. One unique problem of wireless network is the mobility of the nodes, which could lead to the breaking of existing paths and the adding of new links. Handoff problem happens when an aircraft in the network moves from one ground station cell to another cell.

One reasonable way to handle this problem is to divide the network links into two types: stationary links and transient links. Stationary links are those between the stationary nodes or slowly moving nodes, which are likely to exist for a long time. While transient links are those between nodes moving very fast. When an existing path is broken, the traffic flow is rerouted to another feasible path. During the period after the old path is broken and before the new path is set up, best-effort routing is used to route the traffic flow.

For this reason, wireless network normally can only provide *soft QoS*, which means the required QoS is not guaranteed for some transient time periods, when the routing path is broken or the network is partitioned due to the moving of network nodes.

8.5.2.1 Unicast Routing Problem

As we can see from Figure 8.7, if a mobile node sends packages from m to D, Each edge (e.g., a to b in Figure 8.7, or more general, u to v), is represented by the link between two vertices $e = (u, v)$ and has associated q weights corresponding to QoS metrics such that $W_i(u, v) \geq 0$ and $i = 1, 2,\dots, q$. The constraint for each QoS metric is L_i.[13] The multiconstrained path (MCP) problem is to find a path P from a source m to a destination D such that all the QoS constraints are met, as depicted in the following equation:

$$w_i(P) \leq L_i, \quad i = 1, 2,\dots, q$$

The paths that satisfy these constraints are called feasible paths. The solution of the MCP problem requires a path computation algorithm that finds paths that satisfy all the constraints as expressed in the equation. Since the optimal solution of this type of problem for multiple additive and independent metrics is NP-complete, usually heuristics or approximation algorithms are used.

8.5.2.2 Multicast Routing Problem

As we know, the mathematical model of a network is a directed graph $G = (V, E)$ with n nodes ($|V| = n$) and l links ($|E| = l$),[14] where V is a set of nodes and E is a set of links, respectively. Each link $e = (i, j) \in E$ is associated with two parameters, namely, a link cost $c(e) \geq 0$ and a link delay $d(e) \geq 0$. The link cost, $c(e)$, can be associated with the utilization of the link. A higher utilization is represented by a higher link cost. The link delay, $d(e)$, is the sum of the perceived queuing delay, transmission delay, and propagation delay. We define a path as a sequence of links such that $(u, i), (i, j),\dots, (k, v)$, belongs to E.

Let $P(u, v) = \{(u, i), (i, j), \dots, (k, v)\}$ denote the path from node u to node v. If all nodes $u, i, j,\dots, k, v$ are distinct, then we say that it is a simple directed path. We define the length of the path $P(u, v)$, denoted by $n(P(u, v))$, as the number of links in $P(u, v)$. A given source node $s \in V$ and destination node $d \in v$, $(2^{s \to d}, \infty)$ is the set of all possible paths from s to d. $(2^{s \to d}, \infty) = \{P_k(s, d) | \{\text{all possible paths from } s \text{ to } d; s, d \in V, \forall k \in \Lambda\}$, where Λ is an index set. The path cost of P_k is given by $\phi_c(P_k) = \sum_{e \in P_k} c(e)$, and similarly, the path delay of P_k is given by $\phi_D(P_k) = \sum_{e \in P_k} d(e), \forall P_k \in (2^{s \to d}, \infty), (2^{s \to d}, \Lambda)$ and is the set of paths from s to d for which the end-to-end delay is bounded by Δ. Therefore, $(2^{s \to d}, \Delta) \subseteq (2^{s \to d}, \infty)$.

For the multicast communications, messages need to be delivered to all receivers in the set $M \subseteq V_{\setminus \{s\}}$, which is called the multicast group, where $|M| = m$. The path traversed by messages from the source s to a multicast receiver, m_i, is given by $P(s, m_i)$. Thus, the multicast tree can be defined as $T(s, M) = \cup_{m_i \in M} P(s, m_i)$ and the messages are sent from s to M through $T(s, M)$. The tree cost of tree $T(s, M)$ is $\phi_c(T(s, M)) = \sum_{c \in T} c(e)$ and the tree delay is $\phi_D(T(s, M)) = \max(\phi_D(P(s, m_i)) | \forall m_i \in M)$

In order to measure the complexity of different QoS routing algorithms, often called a routing decision problem is NP-complete when it is both in NP and NP-hard. The set of

NP-complete problems is often denoted by NP-C or NPC. NP-complete problems are often addressed using heuristic methods and approximation algorithms.

8.6 QoS Routing Algorithm for Unicast and Multicast

In the previous section we gave the mathematical model of the QoS routing problem that had been carefully reviewed by some researchers.[12] In this section, first some classical algorithms will be discussed, and then we discuss some research on algorithms that was mainly published from 2010 to 2014.

8.6.1 Protocols and Algorithms for Unicast Routing

The unicast information transferred by ATN/IPS can be categorized into two classes, that is, ground–ground application and air–ground application. Only surveillance data (i.e., automatic dependent surveillance, ADS) is multicast; all other applications can be treated as unicast. There are three kinds of routing protocols available to route unicast packets.

1. *Distance vector routing protocol*: Distance vector is a simple routing protocol that bases its routing decision on the number of hops between source and destination. A route with the least number of hops is considered the best route. Every router advertises its best routes to other routers. Ultimately, all routers build up their network topology based on the advertisements of their peer routers. Example: RIP, currently is RIPv2, described in RFC2453.

2. *Link state routing protocol*: A slightly complicated protocol than distance vector. It takes into account the states of links of all the routers in a network. This technique helps routes build a common graph of the entire network. All routers then calculate their best path for routing purposes. Examples: open shortest path first (OSPF) and intermediate system to intermediate system (IS–IS).

3. *Path vector routing protocol*: It will maintain the path information that gets updated dynamically. Updates that have looped through the network and returned to the same node are easily detected and discarded. It is different from the *distance vector routing protocol* and *link state routing protocol*. Each entry in the routing table contains the destination network, the next router, and the path to reach the destination. Example: BGPv4 that was selected in ATN/IPS as the interdomain protocol.

It should be noted that the above classification method of QoS routing is *not* a conflict with the classification method in Section 8.6.1 (source routing/hop-by-hop routing/hierarchical routing). The difference between these two classification methods are the different perspectives of QoS routing. Actually, the research interest of unicast QoS routing is mainly concentrated on *link state routing protocol* that considers more state information, including bandwidth, delay, jitter, and so on.

The most commonly used and classical shortest path algorithms are *Dijkstra's algorithm*, *Bellman–Ford algorithm*, and *Floyd–Warshall algorithm*; many extended algorithms are based on them. The details of these algorithms are carefully discussed in Chapter 2 of Medhi's book.[14] Chen et al.[11] and Curado et al.[12] have given a thorough review of unicast algorithms.

8.6.2 Protocols and Algorithm for Multicast Routing

Some multicast routing protocols that were used to exchange multicast routing information had already been used in a practical system. These protocols are based on the network model and some related algorithms. In this section, some existing multicast protocols will be presented and also the algorithms will be discussed. As the ATN/IPS standard did not specify the intradomain protocol, it is necessary to understand some of these protocols to design and analyze an ATN/IPS network.

We introduce three commonly used multicast routing protocols, namely, DVMRP (distance vector multicast routing protocol), MOSPF (multicast extensions to OSPF) protocol, and PIM-DM (protocol independent multicast-dense mode) protocol. They are very efficient in environments where multicast group members are densely distributed over the network.

1. *DVMRP*: Derived from the RIP protocol, DVMRP is is the oldest routing protocol that supports multicast. The router generates a routing table with the multicast group. The routing table has knowledge of corresponding distances (i.e., number of devices/routers between the router and the destination). When a multicast packet arrives at a router, the router will forward it according to the interfaces specified in the routing table, i.e., delivering tree.

   Delivery trees are constructed according to the information on the previous hop back to the source. First, the router will forward the received packets via each interface except the one at which the packet arrived at. Second, if a router receives the packet transmitted by the source router in the first step, but does not want it to be part of this multicast group, it sends a "prune message" back to the source. These two steps are called reverse flooding technique.

2. *MOSPF*: Defined in RFC 1584 as built on top of OSPF (Version 2, RFC 1583), MOSPF extended the OSPF protocol. The routers that support MOSPF will maintain a table of the network topology. This table is constructed via the unicast OSPF link-state routing protocol. Both MOSFP and DVMRP do not support tunnels.

   First, the source router transmits an OSPF link state information advertisement, then based on the information, the source router can create a distribution tree to identify each multicast source and group, and another tree for active sources sending to the group. The source and all destination group members should reside in the same OSPF area, or the whole autonomous system is a single OSPF area. Note here the route advertisement is realized by Internet Group Management Protocol (IGMP).

3. *PIM-DM*: Developed by the interdomain multicast routing (IDMR) working group of the IETF. PIM-DM requires the existence of a unicast routing protocol to build a routing table. It can be treated as the follower of DVMRP protocol. The multicast date from the source router will "flood" to all parts of the network. The last-hop router will "prune" back to the source if there is no destination group. Dense mode is more efficient when the communication bandwidth is adequate and the group members are densely located among regions.

Some algorithms from simple to complex are discussed as follows:

1. The simplest method is the *flooding* algorithm, which has been already used in protocols such as OSPF, is the simplest technique for delivering the multicast datagram to the routers of an internetwork. When receiving a multicast packet, the router will first check whether it has seen this particular packet before or it is the first time that packet has reached this router. If it is the first time, the router will forward the packet on all interfaces, except the one from which the packet has been received. Otherwise, the router will simply discard the packet. This way we make sure that all routers in the internetwork will receive at least one copy of the packet.

   This type of setup creates unnecessary traffic and wastes precious network resources. Each host machine has to process the packet as it arrives on the interface card, thus, wasting CPU cycles. If the volume of the multicast stream is high enough, it could potentially cause other relevant and control traffic to be dropped.

2. *Spanning tree*: A better algorithm than *flooding* is the *spanning tree* algorithm. This algorithm that has been already used by IEEE-802 MAC bridges is powerful and easy to implement. In this algorithm, a subset of internetwork links is selected to define a tree structure (loop-less graph, it contains no more cycles) such that there is only one active path between any two routers. Since this tree spans to all nodes in the internetwork it is called spanning tree. Whenever a router receives a multicast packet, it forwards the packet on all the links that belong to the spanning tree except the one on which the packet has arrived, guaranteeing that the multicast packet reaches all the routers in the internetwork. Obviously, the only information a router needs to keep is a Boolean variable per network interface, indicating whether the link belongs to the spanning tree or not.

3. The *reverse path broadcasting* (RPB) algorithm: This is a modification of the *spanning tree* algorithm. Instead of building a network-range spanning tree, an implicit spanning tree is constructed for each source. Based on this algorithm, whenever a router receives a multicast packet on link "L" and from source "S", the router will check and see if the link "L" belongs to the shortest path toward S. If this is the case, the packet is forwarded on all links except L. Otherwise, the packet is discarded. Three multicast trees from two sources of the test network are shown in Figure 8.8.

   This algorithm is efficient and easy to implement. Furthermore, since the packets are forwarded through the shortest path from the source to the destination nodes, it is very fast. The RPB algorithm does not need any mechanism to stop the forwarding process. The routers do not need to know about the entire spanning tree and since the packets are delivered through different spanning trees (and not a unique spanning tree), traffic is distributed over multiple tress and network is better utilized. Nevertheless, the RPB algorithm suffer from a major deficiency: it does not take into account the information about multicast group membership for constructing the distribution trees.

   There are other two algorithms that belong to RPB family, TRPB (truncated reverse path broadcasting) and RPM (reverse path multicasting).

In recent published papers, many intelligent algorithms were proposed such as genetic, particle swarm, and simulated annealing algorithm.[17,18] We only cite some papers that published in some relatively important journals.

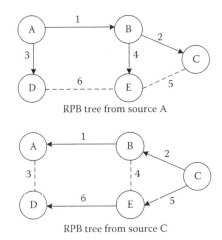

RPB tree from source A

RPB tree from source C

FIGURE 8.8
Reverse path broadcasting algorithm.

Leela[20] introduced a genetic algorithm (GA)–based multicast algorithm that satisfies the multiple constraints of multimedia applications. A heuristic called multiconstraint QoS unicast routing using genetic algorithm (MURUGA), which incorporates multiple constraints required by multimedia applications to find the feasible path satisfying the constraint requirement has been proposed, designed, and simulated. The performance analysis of MURUGA has also been carried out and compared with existing algorithms. The results confirm that MURUGA performs better in terms of time taken to return feasible paths.

Kim[13] studied how to obtain Steiner trees appropriately for efficient multicast routing. Kim first introduces a scheme for generating a new weighted multicast parameter by efficiently combining two independent measures: cost and delay. The proposal method was called weighted parameter for multicast trees (WPMT) algorithm. WPMT can be adjusted by the weight, $\omega \in [0,1]$. For instance, if ω approaches 0, then the delay of the multicast tree may be relatively lower than the delay of other trees that are obtained as ω approaches 1. Otherwise, as the weight approaches 1, the cost of the obtained tree may be relatively lower compared with other trees. A case study also was given to illustrate how to find an appropriate Steiner tree for each ω. The simulation results show that the use of the proposed WPMT produces results similar to the k-minimum Steiner tree algorithm.

Szymanski[19] discussed a multicommodity maximum-flow minimum-cost routing algorithm. The algorithm computes maximum-flow routings for all smooth unicast traffic demands within the capacity region of a network subject to routing cost constraints. The edge cost can be a distance, reliability, congestion, or an energy metric. It is shown that every network has a finite bandwidth–cost capacity. The bandwidth–distance and the bandwidth–energy capacities are explored. The routing algorithm requires the formulation of two linear programs (LPs). The first LP finds a multicommodity maximum flow, when the flows are constrained to a subgraph of the network to enforce cost constraints. The second LP minimizes the routing cost, given that the maximum flow is fixed. A related constrained multicast max-flow–min-cost algorithm is also presented, to maximize the throughput of a multicast tree using network coding, subject to routing cost constraints. These algorithms have polynomial-time solutions, whereas, traditional multipath routing algorithms can be NP-Hard. The addition of routing cost constraints can significantly reduce the size of the LPs resulting in faster solutions, with lower edge utilizations and with higher energy efficiencies. The application of these algorithms to route-aggregated video streams from cloud data centers in a future-Internet network, with improved throughput, energy efficiency, and QoS guarantees is presented.

Sun[21] presented a modified quantum-behaved particle swarm optimization (QPSO) method for QoS multicast routing. Sun discussed that QoS multicast routing is converted into an integer programming problem with QoS constraints and is solved by the QPSO algorithm combined with loop deletion operation. The QPSO-based routing method, along with the routing algorithms based on particle swarm optimization (PSO) and GA, is tested on randomly generated network topologies for the purpose of performance evaluation. The simulation results show the efficiency of the proposed method on the QoS routing problem and its superiority to the methods based on PSO and GA.

8.7 Conclusion

In this chapter, in order to give a more clear and practical reference about how the QoS function is realized in ATN, we combined the ATN environment with IPv6, MIPv6, and NEMO.

Starting from some basic knowledge, we first gave the requirement of QoS in ATN/IPS, then we discussed QoS in two sections, since L2-QoS and L3-QoS are different but are closely related. In Section 8.3, we gave an introduction of the next generation L-DACS1 system, and mainly spoke about some RRM algorithm, and gave some figures to give a clear description. This RRM algorithm is not referred in L-DACS1 specification but is very useful to design a real system. L3-QoS and its application in ATN is introduced in Section 8.4 and as QoS is not defined in mobile environment, some useful research achievements is given in Section 8.4.6.

QoS optimization is highly dependent on the mathematical graph model, hence, we have discussed the unicast and multicast models in Sections 8.5 and 8.6, and we have also introduced some recently published methods for further study. In fact, QoS should be considered not only based on mathematical models but also on practical network structure; how to combine these two in ATN/IPS can be studied in the next step of research.

References

1. ICAO. Manual of technical provisions for the aeronautical telecommunication network (ATN), 9705/AN-956, Second Edition (Effective December 10, 1999), 2002.
2. Eurocontrol/FAA. Action plan 17 future communications study final conclusions and recommendations report, Version 1.1, November 2007.
3. Eurocontrol/FAA. Communications operating concepts and requirements for the future radio system, Eurocontrol/FAA, Technical report, May 2007.
4. Sajatovic, M. et al. L-DACS1 system definition proposal: Deliverable D2. Brussels, Belgium: Eurocontrol, Version 1, 2009.

5. Sajatovic, M. et al. Updated L-DACS1 system specification, SESAR P15 2, 2011. http://www.ldacs.com/publications-and-links/.

6. Ayaz, S. Mobility and radio resource management in future aeronautical mobile networks. PhD thesis. Erlangen, Germany: University Erlangen-Nürnberg, 2012.

7. Shreedhar, M. and Varghese, G. Efficient fair queueing using deficit round robin. *ACM SIGCOMM Computer Communication Review*, 25(4), 231–242, 1995.

8. Lee, C.-W., Chen, M.C., and Sun, Y.S. Protocol and architecture supports for network mobility with QoS-handover for high-velocity vehicles. *Wireless Networks*, 19(5), 811–830, 2013.

9. Kan, Z., Zhang, D., Zhang, R., and Ma, J. QoS in mobile IPv6. In *Proceedings of International Conferences on Info-Tech and Info-Net (ICII 2001)*, Beijing, China, 2001, Vol. 2, pp. 492–497.

10. Seite, P. et al. Quality-of-service option for proxy mobile IPv6. Internet Engineering Task Force. (2014). https://tools.ietf.org/html/rfc7222.

11. Chen, S. and Nahrsted, K., An overview of quality of service routing for next-generation high-speed networks: Problems and solutions. *IEEE Network*, 12(6), 64–79, November/December 1998.

12. Curado, M. and Monteiro, E. A survey of QoS routing algorithms. In *Proceedings of the International Conference on Information Technology (ICIT 2004)*, Istanbul, Turkey, 2004.

13. Kim, M., Choo, H., Mutka, M.W., Lim, H.J., and Park, K. On QoS multicast routing algorithms using k-minimum Steiner trees. *Information Sciences*, 238, 190–204, 2013.

14. Medhi, D. *Network Routing: Algorithms, Protocols, and Architectures*. Boston, MA: Morgan Kaufmann, 2010.

15. ICAO. Aeronautical Communications Panel (Working Group I), *Manual for the ATN Using IPS Standards and Protocols*, Doc 9896, First Edition, April 2014. http://www.icao.int/safety/acp/Pages/.

16. Brandes, S., Epple, U., Gligorevic, S., Schnell, M., Haindl, B., and Sajatovic, M. Physical layer specification of the L-band digital aeronautical communications system (L-DACS1). In *Integrated Communications, Navigation and Surveillance Conference*, Arlington, VA, 2009, p. 1.

17. Yin, P.Y., Chang, R.I., Chao, C.C., and Chu, Y.T. Niched ant colony optimization with colony guides for QoS multicast routing. *Journal of Network and Computer Applications*, 40, 61–72, 2014.

18. Lu, T. and Zhu, J. Genetic algorithm for energy-efficient QoS multicast routing. *IEEE Communications Letters*, 17(1), 31–34, 2013.

19. Szymanski, T.H. Max-flow min-cost routing in a future-internet with improved QoS guarantees. *IEEE Transactions on Communications*, 61(4), 1485–1497, 2013.

20. Leela, R., Thanulekshmi, N., and Selvakumar, S. Multi-constraint QoS unicast routing using genetic algorithm (MURUGA). *Applied Soft Computing*, 11(2), 1753–1761, March 2011.

21. Sun, J., Fang, W., Wu, X., Xie, Z., and Xu, W. QoS multicast routing using a quantum-behaved particle swarm optimization algorithm. *Engineering Applications of Artificial Intelligence*, 24(1), 123–131, 2011.

9

Time Division Multiplexing in Satellite Aeronautical Communications System

Sarhan M. Musa and Zhijun Wu

CONTENTS

9.1 Introduction

Aeronautical communications has advanced different solution systems that allow aircraft to maintain a link with the ground while in flight due to the growth of worldwide air traffic and the increased need for communication safety. Today, we observe that satellite solution systems are effective in supporting aeronautical communications. Indeed, satellite communications is currently playing a major role toward the implementation of a global communication infrastructure, especially given the explosive growth of wireless technology. This chapter presents the analysis and simulation of blocking and clipping probabilities for time division multiplexing (TDM) in a satellite aeronautical communications system. Specifically, we illustrate the evaluation of multiplexing systems in which the number of input sources is greater than the number of available channels. For the case of the blocking situation in synchronous TDM, we investigate the blocking probability and the average number of busy channels that can be delivered. For the case of clipping in statistical TDM, we examine the clipping probability and the expected number of busy channels that can be delivered. We compare the blocking and clipping probabilities for a fixed number of sources and different numbers of channels. We also compare the expected number of busy available channels for synchronous TDM and the average number of busy channels for statistical TDM methods for a fixed number of sources and different numbers of channels.

Satellite communications was first deployed in the 1960s for military applications. Satellites have played an important role in both domestic and international communications networks since the launching of the first commercial communication satellite by NASA in 1965. They have brought voice, video, and data communications to areas of the world that are not accessible with terrestrial lines. By extending communications to the remotest parts of the world, virtually everyone can be part of the global economy.

An aircraft can be connected to the ground via a variety of different wireless access technologies. A list of these technologies and their characteristics is provided in [1].

Satellite communications is not a replacement of existing terrestrial systems but rather an extension of wireless systems. However, satellite communications has the following merits over terrestrial communications [2]:

1. *Coverage*: Satellites can cover a much larger geographical area than the traditional ground-based system. They have a unique ability to cover the globe.

2. *High bandwidth*: A Ka-band (27–40 GHz) can deliver throughput of gigabits per second rate.

3. *Low cost*: A satellite communications system is relatively inexpensive because there are no cable-laying costs and one satellite covers a large area.

4. *Wireless communication*: Users can enjoy untethered mobile communication anywhere within the satellite coverage area.

5. *Simple topology*: Satellite networks have simpler topology, which results in more manageable network performance.

6. *Broadcast/multicast*: Satellites are naturally attractive for broadcast/multicast applications.

7. *Maintenance*: A typical satellite is designed to be unattended, requiring only minimal attention by customer personnel.

8. *Immunity*: A satellite system will not suffer from disasters such as floods, hurricanes, fire, and earthquakes and will therefore be available as an emergency service should terrestrial services be knocked out.

Based on the location of the orbit, satellites are divided into three categories: geostationary Earth orbit (GEO), medium Earth orbit (MEO), and low Earth orbit (LEO) as shown in Figure 9.1. A comparison of the three satellite types is given in Table 9.1.

In order to facilitate satellite communications and eliminate interference between different systems, international organizations govern the use of satellite frequency. The International Telecommunication Union is responsible for allocating frequencies to satellite services.

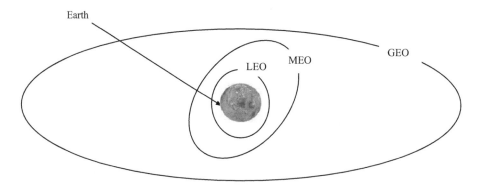

FIGURE 9.1
The three common types of satellites: geostationary Earth orbit, medium Earth orbit, and low Earth orbit.

TABLE 9.1

Comparison of Geostationary Earth Orbit, Medium Earth Orbit, and Low Earth Orbit

Type (km)	Altitude Coverage	Advantages	Disadvantages
LEO 300–1,000	Spot	Low path loss	Less coverage
		High data rate	Need many satellites
		Low delay	Short orbital life
		Low launch cost	High Doppler
		Less fuel	Highly complex
MEO 1,000–10,000	Region	Moderate path loss	Multiple satellites
		Moderate launch cost	Moderate coverage
		Less fuel	Highly complex
GEO 36,000	Earth	Global coverage	High path loss
		Need few satellites	Long delay
		Long orbital life	Low data rate
		Low Doppler	High launch cost
		Less complex	Fuel for station keeping

TABLE 9.2

Satellite Frequency Allocations

Frequency Band	Range (GHz)
L	1–2
S	2–4
C	4–8
X	8–12
Ku	12–18
K	18–27
Ka	27–40

The frequency spectrum allocations for satellite services are given in Table 9.2. In fact, the signals between the satellite and the Earth stations travel along line-of-sight paths and experience free-space loss that increases as the square of the distance.

Link budget of satellite is calculated for both the uplink from the gateway to the satellite and the downlink from the satellite to the user terminal as shown in Figure 9.2. Link budget of a satellite can be computed by the single link equation as

$$\frac{C}{N_0} = \left(P_T G_T\right)\left(\frac{1}{L}\right)\left(\frac{G_R}{T}\right)\left(\frac{1}{K}\right) \tag{9.1}$$

where
C/N_0 is the signal to noise power ratio in dBHz
$P_T G_T$ is the transmitter equivalent isotropic radiated power in dBW
P_T is the carrier power
G_T is the transmit antenna gain
$1/L$ is the cumulative path loss
G_R is the antenna gain of the receiver
T is the receiver system temperature
K is the Boltzmann constant

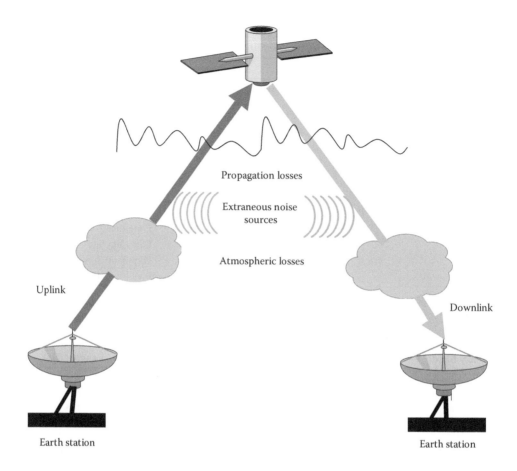

FIGURE 9.2
(See color insert.) Satellite link budget.

A satellite band is divided into a number of separated portions: one for Earth-to-space links (the uplink) and one for space-to-Earth links (the downlink). Separate frequencies are assigned for the uplink and the downlink. Table 9.3 provides the general frequency assignments for uplink and downlink satellite frequencies. We notice from the table that the uplink frequency bands are slightly higher than the corresponding downlink frequency band. This is to take advantage of the fact that it is easier to generate RF power within a ground station than it is on board a satellite. In order to direct the uplink transmission to

TABLE 9.3

Typical Uplink and Downlink Satellite Frequencies

Uplink Frequencies (GHz)	Downlink Frequencies (GHz)
5.925–6.425	3.700–4.200
7.900–8.400	7.250–7.750
14.00–14.50	11.70–12.20
27.50–30.0	15.70–20.20

a specific satellite, the uplink radio beams are highly focused. In the same way, the downlink transmission is focused on a particular *footprint* or area of coverage.

Multiplexing is the concept of allowing the simultaneous transmission of multiple signals across a single data link. Frequency division multiplexing (FDM), wavelength division multiplexing (WDM), and TDM are multiplexing techniques.

FDM is an analog technique that can be applied when the bandwidth of a link (in hertz) is greater than the combined bandwidths of the signals to be transmitted. In FDM, signals generated by each sending device modulate different carrier frequencies. These modulated signals are then combined into a single composite signal that can be transported by a link.

WDM is a light technique that can be designed to use the high-data-rate capability of fiber-optic cable. The optical fiber data rate is higher than the data rate of metallic transmission cable, but using a fiber-optic cable for a single line wastes the available bandwidth. Multiplexing allows us to combine several lines into one.

TDM is a digital process that allows several connections to share the high bandwidth of a link. Instead of sharing a portion of the bandwidth as in FDM, time is shared. Each connection occupies a portion of time in the link.

Multiple access technologies allow different users to utilize the satellite's resources of power and bandwidth without interfering with each other. Satellite communications systems use different types of multiple access technology, including frequency division multiple access (FDMA), time division multiple access (TDMA), and code division multiple access (CDMA). The access technology can vary between the uplink and downlink channels.

The ability of multiple Earth stations or users to access the same channel is known as FDMA. In FDMA, each user signal is assigned a specific frequency channel. One disadvantage of FDMA is that once a frequency is assigned to a user, the frequency cannot be adjusted easily or rapidly to other users when it is idle. The potential for interference from adjacent channels is another major shortcoming.

In TDMA, each user signal is allotted a time slot. A time slot is allocated for each periodic transmission from the sender to a receiver. The entire bandwidth (frequency) is available during the time slot. This access scheme provides priority to users with more traffic to transmit by assigning those users more time slots than it assigns to low-priority users. Satellite providers will extend the capability and will employ multiple frequency TDMA. If there are N frequencies, each offering M Mbps of bandwidth, then the total available bandwidth during a time slot is N × M Mbps. Although FDMA techniques are more commonly employed in satellite communications systems, TDMA techniques are more complex and are increasingly becoming the de facto standard.

In CDMA, users occupy the same bandwidth but use spread spectrum signals with orthogonal signaling codes. This technique increases the channel bandwidth of the signal and makes it less vulnerable to interference. CDMA operates in three modes: direct sequence, frequency hopping, and time hopping.

In recent years, significant effort has been made toward evaluating blocking probabilities experienced by customers contending for a commonly shared resource. By definition, the blocking probability is the probability that a connection service request is denied due to insufficient network resources. Proportional differentiation models have been proposed as effective methods for scalable differentiation service provision into optical WDM networks with blocking probability to various traffic classes [3–5].

Once a channel is assigned to a given talkspurt, the channel is held till the spurt ends. The occurrence of freeze-out typically causes the initial part of a talkspurt to be clipped. If a talkspurt sees no channel available upon its arrival, the initial portion of the talkspurt will be clipped until a newly freed channel can be assigned. Clipping probability is used in measuring video quality [6–8].

In this work, we will illustrate the multiplexing models used in satellite communications systems, especially TDM, focusing on synchronous TDM and statistical TDM methods.

9.2 Multiplexing Models and Results

The technology of satellite aeronautical communications can support fixed and wireless data, voice, and video communications, Internet connections, and enterprise networking. Satellite communications successfully use a continuously transmitted signal of TDM for the outbound (downlink) to improve of transmission quality instead of orthogonal FDM, because of the linearity requirement on the power amplifier. TDMs are used in satellite networks for maximum transmission capacity of a high-bandwidth line. Multiplexing allows many communication sources to transmit data over a single physical line.

There are a number of multiplexing methods used in satellite communications systems. One of the commonly used methods in such systems is the TDM technology as in [9–12]. Blocking and statistical probabilities are applied for TDM multiplexing as in [13–22]. In this chapter, we will analyze and simulate the blocking and statistical probabilities of TDM applied to satellite communications systems.

Figure 9.3 shows a satellite communications system using TDM technology in interactive and sending/receiving different applications based on the importance of response time. TDM is used in the outbound link between the source (sender or host) and the user (receiver). A TDM system is a high-speed data stream scheme that acts at layer 1 (physical layer) of the Open Systems Interconnection model and at layer 4 (network interface) of the Transmission Control Protocol over Internet Protocol model. In TDM technology, users take turns in a predefined way, each one periodically getting the entire bandwidth

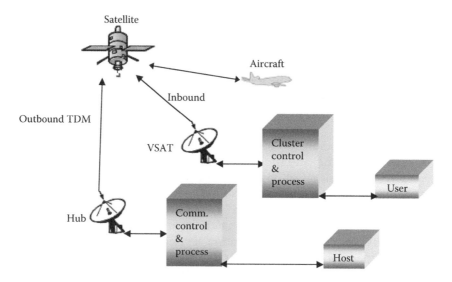

FIGURE 9.3
Satellite communications system with TDM.

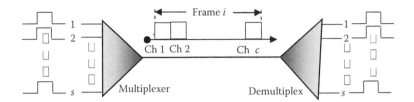

FIGURE 9.4
A time division multiplexer with s inputs and c channels.

for a portion of the total scanning time. The input source s is divided into frames, and each frame is subdivided into time slots (channels), c, where each channel is allocated to one input as shown in Figure 9.4. Packets arrive on s lines, and the multiplexer scans them, forming a frame with c channels on its outgoing link. There are two different types of TDMs to deal with the different ways in which channels of frames could be allocated as synchronous and statistical (asynchronous) and it has been used with other techniques as solution for satellite communications networks such as TDMA, FDMA, and PAMA.

9.2.1 Synchronous TDM

In synchronous TDM, a frame is divided in fixed-sized channels and channels are allocated to input sources in a fixed way. The Quality of Service (QoS) of synchronous TDM is based on how its transmission system is set up. For example, the multiplexer is inefficient when the number of users is greater than the available channels. This is true since the multiplexer scans all input source lines without exceptions and the scanning time for each input source line (each connected to a user) is reallocated, as well as this time for a particular input source line is not altered by the system control. The scanner should stay on that input source line, whether or not there are data for scanning within that time slot. A synchronous TDM can also be programmed to produce same-sized frames; the lack of data in any channel potentially creates changes to average bit rate on the ongoing link.

To analyze a synchronous multiplexer, let t_a and t_d be the mean time for active input source and the mean time for idle input source, respectively. Let us assume that the values of t_a and t_d are random and exponentially distributed. (This assumption is based on experience.) Also, consider a TDM with a number of requesting input sources, s, greater than available channels, c, where $s > c$; the TDM will react by blocking. The unassigned input sources are not transmitted and therefore remain inactive. The probability that an input source is active, ρ, can be obtained by $\rho = t_a/(t_a + t_d)$.

Let $P_s(j)$ be the probability of j different inputs out of s that are active:

$$P_s(j) = \binom{s}{j}\rho^j\left(1-\rho\right)^{s-j}, \tag{9.2}$$

where $1 \leq j \leq s$. We know that $\sum_{j=0}^{c} P_s(j)$ can never be equal to 1, and in fact we must have $\sum_{j=0}^{s} P_s(j) = 1$. This can lead to normalization of $P_s(j)$ over c available channels. Thus, the probability of j different output of c available channels, $P_c(j)$, is

$$P_c(j) = \frac{\left(\dfrac{s}{j}\right)\left(\dfrac{\rho}{1-\rho}\right)^j}{\displaystyle\sum_{i=0}^{c}\left(\dfrac{s}{j}\right)\left(\dfrac{\rho}{1-\rho}\right)^i}, \tag{9.3}$$

where $0 \le j \le c$ and $0 \le i \le c$. The blocking probability $P_s(c)$ can be obtained when $j = c$:

$$P_s(c) = \frac{\left(\dfrac{s}{c}\right)\left(\dfrac{t_a}{t_d}\right)^c}{\displaystyle\sum_{i=0}^{c}\left(\dfrac{s}{i}\right)\left(\dfrac{t_a}{t_d}\right)^i}, \tag{9.4}$$

where in general $0 \le i \le c$.

Figure 9.5 shows the blocking probability for the fixed number of sources ($s = 10$) and different numbers of channels ($c = 1$, $c = 4$, and $c = 7$), whereas Figure 9.6 shows the blocking probability for the fixed number of sources ($s = 10$) and different numbers of channels ($c = 2$, $c = 5$, and $c = 8$). The blocking probability clearly rises with the increased utilization, ρ, for all three cases; and also it is higher when a fewer number of channels, c, are available.

Then, we can calculate the expected number of busy channels for the multiplexer, $E_c(b)$, by

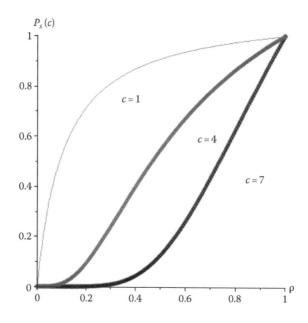

FIGURE 9.5
Comparison of blocking probability, $P_s(c)$, versus probability of active input source, ρ, with $0 \le \rho \le 1$, for $s = 10$, $c = 1$, 4, and 7.

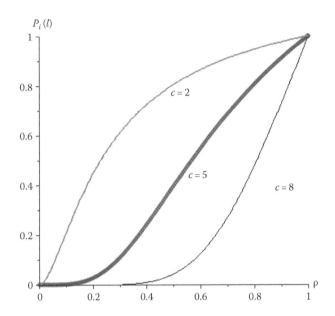

FIGURE 9.6
Comparison of blocking probability, $P_s(c)$, versus probability of active input source, ρ, with $0 \le \rho \le 1$, for $s = 10$, $c = 2, 5,$ and 8.

$$E_c(b) = \frac{\sum_{j=1}^{c} j \left(\dfrac{s}{j}\right)\left(\dfrac{t_a}{t_d}\right)^j}{\sum_{i=0}^{c} \left(\dfrac{s}{i}\right)\left(\dfrac{t_a}{t_d}\right)^i} , \tag{9.5}$$

where $1 \le j \le c$, $0 \le i \le c$, and $\left(\dfrac{\rho}{1-\rho}\right) = \left(\dfrac{t_a}{t_d}\right)$.

Figure 9.7 shows the expected number of busy (unavailable) channels for fixed number of sources ($s = 10$) with different numbers of channels ($c = 1$, $c = 4$, and $c = 7$), while Figure 9.8 shows the expected number of busy (unavailable) channels for fixed number of sources ($s = 10$) with different numbers of channels ($c = 2$, $c = 5$, and $c = 8$). The expected number of busy channels varies in its maximum values based on the interval of utilization.

9.2.2 Statistical TDM

Statistical TDM has high efficiency because a frame's time slots are dynamically allocated, based on demand, and it removes all the empty slots on a frame. But it is difficult to give a guarantee QoS, because of the requirement that additional overhead be attached to each outgoing channel. These additional data are needed because each channel must carry information about which input source line it belongs to. The frame length is available not only because of different channel sizes but also because of the possible absence of some channels.

We consider t_a and t_d as random and exponentially distributed. Also, consider a TDM with a number of requesting input sources, s, greater than available channels, c, where $s > c$; the TDM will react by clipping; the unassigned input sources are partially transmitted. If

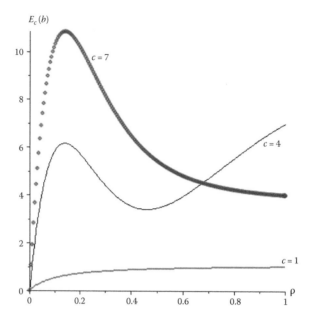

FIGURE 9.7
Comparison on expected numbers of busy available channels, $E_c(b)$, versus probability of active input source, ρ, where $0 \leq \rho \leq 1$, for $s = 10$, $c = 1$, 4, and 7.

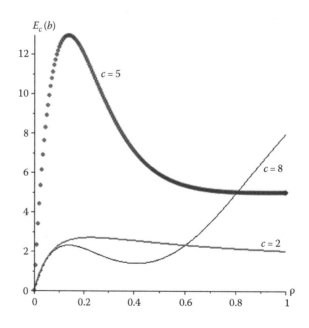

FIGURE 9.8
Comparison on expected numbers of busy available channels, $E_c(b)$, versus probability of active input source, ρ, where $0 \leq \rho \leq 1$, for $s = 10$, $c = 2$, 5, and 8.

more than c input channels are active, we can dynamically choose c out of s active sources and temporarily block other sources. In this temporary blocking, the source is forced to clip or lose data for a short period of time, where the amount of data lost depends on t_a, t_d, s, and c, but the source may return to a scanning scenario if a channel becomes free.

The clipping probability, $P_i(l)$, or the probability that an idle source finds at least c channels busy at the time it becomes active can be calculated by considering all s sources minus 1 (the examining source):

$$P_i(l) = \sum_{i=c}^{s-1} \binom{s-1}{i} \rho^i (1-\rho)^{s-1-i}, \tag{9.6}$$

where $c \le i \le s - 1$.

Figure 9.9 shows the clipping probability for a fixed number of sources ($s = 10$) and different numbers of channels ($c = 1$, $c = 4$, and $c = 7$), while Figure 9.10 shows the clipping probability for a fixed number of sources ($s = 10$) and different numbers of channels ($c = 2$, $c = 5$, and $c = 8$). The clipping probability of two channels has the highest clipping probability compared to five and eight channels.

Clearly, the average number of used channels, $A_c(u)$, is

$$A_c(u) = \frac{\sum_{j=1}^{c} j \binom{s}{j} \left(\frac{t_a}{t_d}\right)^j}{\sum_{i=0}^{c} \binom{s}{j} \left(\frac{t_a}{t_d}\right)^i}, \tag{9.7}$$

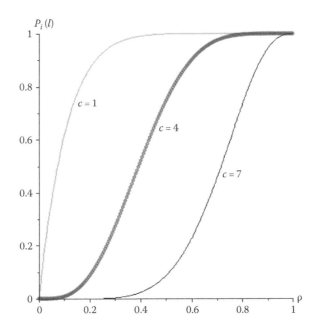

FIGURE 9.9
Comparison of clipping probability, $P_i(l)$, versus probability of active input source, ρ, where $0 \le \rho \le 1$, for $s = 10$, $c = 1, 4,$ and 7.

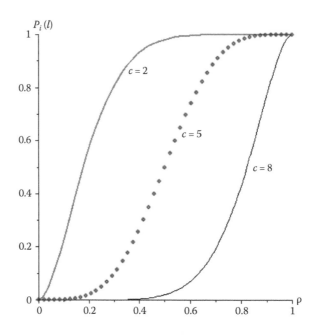

FIGURE 9.10
Comparison of clipping probability, $P_i(l)$, versus probability of active input source, ρ, where $0 \leq \rho \leq 1$, for $s = 10$, $c = 2, 5,$ and 8.

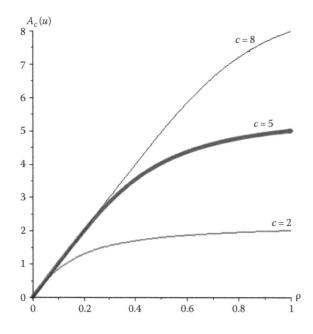

FIGURE 9.11
Comparison of average number of used channels, $A_c(u)$, versus probability of active input source, ρ, where $0 \leq \rho \leq 1$, for $s = 10$, $c = 2, 5,$ and 8.

where $0 \leq i \leq c$ and $1 \leq j \leq c$. Figure 9.11 shows the average number of used channels for a fixed number of sources ($s = 10$) and different numbers of channels ($c = 2$, $c = 5$, and $c = 8$). The average number of used channels of eight channels has the highest average number of used channels compared to the ones for two and five channels.

The average number of busy channels, $A_c(b)$, is

$$A_c(b) = \frac{\sum_{j=1}^{c} j \left(\frac{s}{j}\right)\left(\frac{t_a}{t_d}\right)^j}{\sum_{i=0}^{c} \left(\frac{s}{j}\right)\left(\frac{t_a}{t_d}\right)^i} + c \sum_{j=c+1}^{s} \left(\frac{s}{j}\right)\rho^j(1-\rho)^{s-j}, \tag{9.8}$$

where $0 \leq i \leq c$, $1 \leq j \leq c$, and $c + 1 \leq j \leq s$.

Figure 9.12 shows the average number of busy channels for a fixed number of sources ($s = 10$) and different numbers of channels ($c = 2$, $c = 5$, and $c = 8$). The average number of busy channels for all cases is almost the same up to $\rho = 0.25$, but it differs for $\rho > 0.25$.

Figures 9.13 through 9.15 show the comparison between blocking and clipping probabilities for a fixed number of sources ($s = 10$) and different numbers of channels ($c = 2$, $c = 5$, and $c = 8$, respectively). We observe that the blocking probability is greater than the clipping probability for 2 and 5, but it varies at channel 8.

Figures 9.16 through 9.18 show the expected number of busy available channel for synchronous TDM and the average number of busy channel for statistical TDM methods for a fixed number of sources ($s = 10$) and different numbers of channels ($c = 2$, $c = 5$, and $c = 8$, respectively). We observe that the average number of channels and the expected number of busy available channels vary in the different utilization numbers.

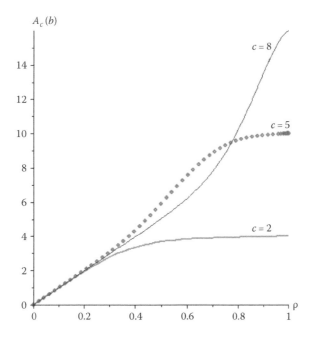

FIGURE 9.12

Comparison of average number of busy channels, $A_c(b)$, versus probability of active input source, ρ, where $0 \leq \rho \leq 1$, for $s = 10$, $c = 2$, 5, and 8.

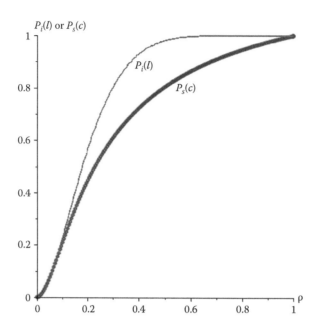

FIGURE 9.13
Comparison of blocking probability, $P_s(c)$, and clipping probability, $P_i(l)$, versus probability of active input source, ρ, with $0 \leq \rho \leq 1$, for $s = 10$, $c = 2$.

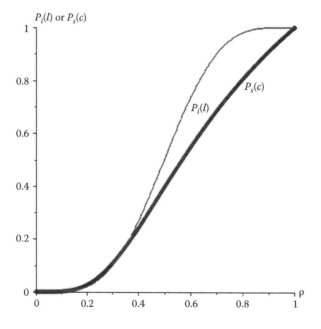

FIGURE 9.14
Comparison of blocking probability, $P_s(c)$, and clipping probability, $P_i(l)$, versus probability of active input source, ρ, with $0 \leq \rho \leq 1$, for $s = 10$, $c = 5$.

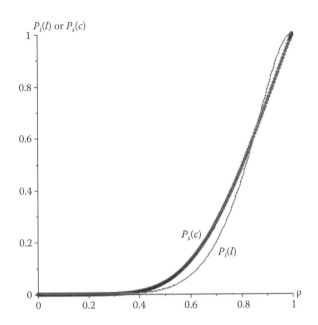

FIGURE 9.15
Comparison of blocking probability, $P_s(c)$, and clipping probability, $P_i(l)$, versus probability of active input source, ρ, with $0 \leq \rho \leq 1$, for $s = 10$, $c = 8$.

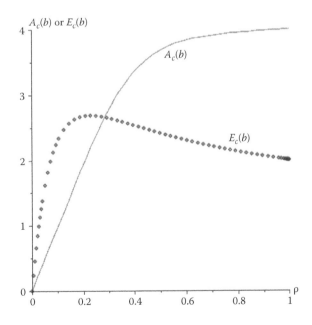

FIGURE 9.16
Comparison of expected number of busy available channels, $E_c(b)$, and average number of busy channels, $A_c(b)$, versus probability of active input source, ρ, where $0 \leq \rho \leq 1$, for $s = 10$, $c = 2$.

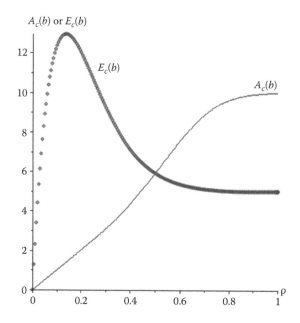

FIGURE 9.17
Comparison of expected number of busy available channels, $E_c(b)$, and average number of busy channels, $A_c(b)$, versus probability of active input source, ρ, where $0 \leq \rho \leq 1$, for $s = 10$, $c = 5$.

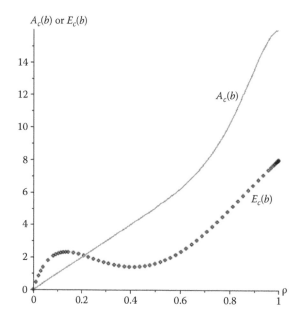

FIGURE 9.18
Comparison of expected number of busy available channels, $E_c(b)$, and average number of busy channels, $A_c(b)$, versus probability of active input source, ρ, where $0 \leq \rho \leq 1$, for $s = 10$, $c = 8$.

9.3 Conclusion

We presented the analysis of TDM applied to satellite aeronautical communications systems when input sources are greater than available channels. The analysis of blocking and clipping probabilities for TDMs was successfully achieved, and results of the analysis were generated. For blocking cases in synchronous TDMs, we illustrated the blocking probability and the average number of busy channels that could be delivered. For the clipping in statistical TDM, we examined the clipping probability and the expected number of busy channels. We compared the blocking and clipping probabilities for a fixed number of sources and different numbers of channels. We also compared the expected number of busy available channel for synchronous TDM and the average number of busy channels for statistical TDM methods for a fixed number of sources and different numbers of channels.

References

1. Airlines Electronic Engineering Committee (AEEC). ARINC Report 811: Commercial aircraft information security concepts of operation and process framework, Aeronautical Radio Inc., p. 138, December 2005.
2. M. N. O. Sadiku, *Optical and Wireless Communications: Next Generation Networks.* Boca Raton, FL: CRC Press, pp. 185–217, 2002.
3. Y. Chen, M. Hamdi, and D. H. K. Tsang, Proportional QoS over WDM networks: Blocking probability, *Proceedings of Sixth IEEE Symposium on Computers and Communications*, pp. 210–215, Hammamet, July 2001.
4. A. Bianco, G. Galante, and M. Mellia, Analysis of call blocking probability in TDM/WDM networks with transparency constraint, *IEEE Communications Letters*, 4(3), 104–106, March 2000.
5. Y. C. Huei and P. H. Keng, A shared time-slot router architecture for TDM wavelength optical WDM, *Proceedings of IEEE Seventh Malaysia International Conference on Communication*, Vol. 2, pp. 672–677, Malaysia, November 2005.
6. C. J. Weinstein, Fractional speech loss and talker activity model for TASI and for packet-switched speech, *IEEE Transactions on Communications*, 26, 1253–1257, August 1978.
7. S. Q. Li, A new performance measurement for voice transmission in burst and packet switching, *IEEE Transactions on Communications*, 35(10), 1083–1094, October 1987.
8. A. Gatherer and M. Polley, Controlling clipping probability in DMT transmission, *Proceedings of the 31st Asilomar Conference on Signals, Systems and Computers*, Vol. 1, pp. 578–584, Pacific Grove, CA, November 1997.
9. O. R. Ramires, Use of TDM/PAMA satellite based communication as an option for the aeronautical communication, *IEEE National Aerospace and Electronics Conference*, pp. 683–686, Dayton, OH, October 2000.
10. Y. Chaehag, TDM framing for gap filler operation in satellite digital multimedia broadcasting System A, *IEEE 59th Vehicular Technology Conference*, Vol. 5, pp. 2782–2786 May 2004.
11. K. Zhang and S. Hryckiewicz, An integrated approach for IP networking over the wideband Gapfiller satellite, *IEEE on Military Communications Conference (MILCOM)*, Vol. 3, pp. 1556–1561, November 2004.
12. K. Eng, and A. Acampora, Fundamental conditions governing TDM switching assignments in terrestrial and satellite networks, *IEEE Transactions on Communications*, 35(7), 755–761, July 1987.

13. W. K. Lai, Y. J. Jin, H. W. Chen, and C. Y. Pan, Channel assignment for initial and handoff calls to improve the call-completion probability, *IEEE Transactions on Vehicular Technology*, 52(4), 876–890, July 2003.

14. J. W. Chong; B. C. Jung, and D. K. Sung, Statistical multiplexing-based hybrid FH-OFDMA system for OFDM-based UWB indoor radio access networks, *IEEE Transactions on Microwave Theory and Techniques*, 54(4), 1793–1801, June 2006.

15. A. I. Elwalid and D. Mitra, Statistical multiplexing with loss priorities in rate-based congestion control of high-speed networks, *IEEE Transactions on Communications*, 42(11), 2989–3002, November 1994.

16. M. Balakrishnan, R. Cohen, E. Fert, and G. Keesman, Benefits of statistical multiplexing in multi-program broadcasting, *International Broadcasting Convention*, 1, 560–565, September 1997.

17. J. Liebeherr, S. Patek, and E. Yilmaz, Tradeoffs in designing networks with end-to-end statistical QoS guarantees, *Eighth International Workshop on Quality of Service (IWQOS)*, pp. 221–230, Pittsburgh, PA, June 2000.

18. Z. L. Zhang, J. Kurose, J. D. Salehi, and D. Towsley, Smoothing, statistical multiplexing, and call admission control for stored video, *IEEE Journal on Selected Areas in Communications*, 15(6), 1148–1166, August 1997.

19. W. C. La and S. Q. Li, Statistical multiplexing and buffer sharing in multimedia high-speed networks: A frequency-domain perspective, *IEEE/ACM Transactions on Networking*, 5(3), 382–396, June 1997.

20. A. Iera, A. Molinaro, S. Marano, and D. Mignolo, Statistical multiplexing of heterogeneous traffic classes in ATM-satellite based networks, *IEEE Wireless Communications and Networking Conference (WCNC)*, Vol. 1, pp. 174–178, New Orleans, LA, September 1999.

21. L. Tancevski and I. Andonovic, Block multiplexing codes for incoherent asynchronous all-optical CDMA using ladder network correlators, *IEE Proceedings on Optoelectronics*, 142(3), 125–131, June 1995.

22. M. Zafer and E. Modiano, Blocking probability and channel assignment in wireless networks, *IEEE Transactions on Wireless Communications*, 5(4), 869–879, April 2006.

10

ATN Transmission Control Algorithm Based on Service-Oriented Architecture

Haitao Zhang

CONTENTS

10.1 Features and Composition of ATN Network

The Aeronautical Telecommunications Network (ATN) is an important part of the communications, navigation, and surveillance/air traffic management system (CNS/ATM) and the basis for a new generation of navigation systems. ATN adopts seven communication models formulated by the International Organization for Standardization (ISO).

10.1.1 Composition of ATN

ATN consists of three subnets: ground–ground communication subnet, air–ground communication subnet, and air–air communication subnet. Currently, the ground/air communication subnet of ATN is mainly made up of very high frequency (VHF) data link, aeronautical mobile satellite communications (AMSS), and secondary surveillance radar mode S data link [1].

1. *VHF data link*: The VHF system works in 118–136 MHz frequency band and its communication coverage radius is inside 200 nautical miles. There are four modes, mode 1, mode 2, mode 3, and mode 4 in the VHF system, and mode 2 and mode 4 have been recognized as an international standard by ICAO, which are also very promising technologies. VHF mode 2 adopts D8PSK modulation mode; each channel bandwidth is 25 kHz, the bit rate is 31.5 kbps, and p-persistent carrier sense shares channel access whose probability p is 13/256. VHF mode 2 can support voice and data communications, which is now widely applied in air–ground communication systems of domestic and international civil aviation.

2. *AMSS*: As a powerful means of communication for aircraft flying over oceans, deserts, Antarctic, Arctic, and other areas where low altitude communication infrastructure is weak, AMSS can provide voice and data communications for aircraft in areas where VHF and HF cannot cover. Currently, INMARSAT is able to provide satellite communications services worldwide, and its satellite beam can cover most of the world's busiest routes, providing services for many users through ground stations and ground earth stations (GESs) and providing communication with channel rates of 9600, 64, and 256 kbps for the users.

3. *Secondary surveillance radar mode S*: Secondary surveillance radar A or C mode coding is limited in number and the exchangeable information (ID, height) is little. All aircraft will receive inquiry signals and respond to them at the same time, which results in mutual signal interference and loss. Mode S is compatible with modes A/C, and the first 24 bits of each inquiry signal serves as an address code of the aircraft, so the identifiable number of aircraft can reach 224, which is more than 4000 times as the current A mode does, being enough to implement an address for an airplane.

10.1.2 Data Information and Related Systems of ATN

Information transmitted between the aircraft and the ground includes air traffic management (ATM) information, airlines operation information, and personal communication information of travelers by air. According to the priority, the ATM information takes precedence over airlines operation information and airlines operation information takes precedence over personal communication information. The ATM information itself also includes various types of messages. ATM information in air-to-ground communication includes automatic dependent surveillance messages, navigational intelligence messages, directory services messages, and the pilot–controller data link communication messages. These messages also have different QoS requirements. ICAO divided air traffic management information into eight types from A to H according to different transmission delay requirements, where A type is reserved and B to H are applied to different air–ground messages, as is shown in Table 10.1.

1. *Automatic dependent surveillance (ADS)*: Automatic dependent surveillance-broadcast (ADS-B) is an important part of a new generation of air traffic management systems, as ADS-B, compared with conventional radar, provides higher security, costs less, and is superior in operation. Aircraft equipped with ADS-B can automatically keep sending the ADS information, including the aircraft's identification code, latitude, longitude, velocity, direction, altitude, time, and other auxiliary information to the ground controller's ADS terminal or other aircraft. The current situation in the air can also be shown on a display terminal on the ground and other aircraft ground controllers can promptly detect the introduction errors of waypoints and ATC loop error, and carry out compliance overseeing for the current flight plan and deviation detection based on ADS information; this will help find out the deviation of the release path of the aircraft flight path and provide better continuity for the path and track position, velocity, and acceleration estimation to prevent collision, which in turn ensures flight safety intervals and reduces the number of false alarms.

2. *Controller–pilot data link communications (CPDLC)*: CPDLC enables pilots and controllers to carry out two-way communication directly via a data link so that they can send a variety of requests and control commands. These requests and commands

TABLE 10.1

Transmission Delay Requirements of ATM Information

One-Way Transmission Delay in ATN (Success Rate of Transmission Is 95%, and Unit Is Second)	Types of ATM Information
Reserved	A
4.5	B
7.2	C
13.5	D
18	E
27	F
50	G
100	H

are implemented as data messages, thus avoiding ambiguity caused by direct voice communication. Using the data link can increase system capacity and reduce the workload of controllers and pilots.

3. *Flight information system* (*FIS*): This service supports pilots with all sorts of information during flight. At present, information is transmitted through voice with a small amount of information and a long time of transmission. There are two ways of using the data link: broadcast and request type. The information would be stored in the database after the aircraft receives it for pilots to check. FIS can provide a variety of text or graphical information on weather, airport, and aircraft equipment.

10.1.3 Structure of ATN Network System

The ATN network system is divided into end system (ES) and intermediate system (IS), as shown in Figure 10.1. The end system contains each user's computer unit in the ATN, and

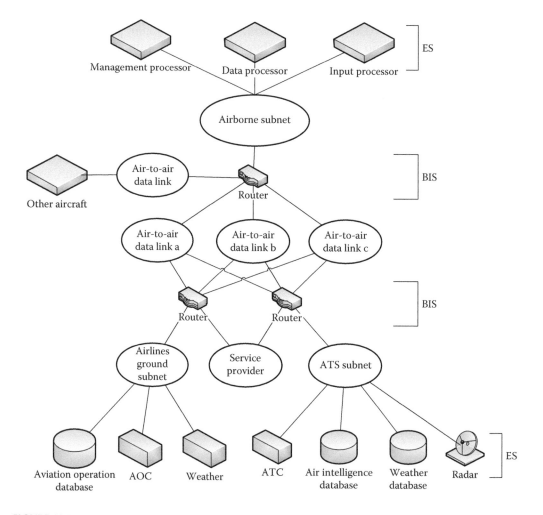

FIGURE 10.1
Structure of ATN.

provides end-to-end communications services for upper applications, and also provides user interface for man–machine interaction. Airborne ES includes display and control systems in aircraft cockpit and operator's workbenches in aircraft task electronic systems such as radar, command guidance, and other workbenches. Ground ES includes all types of end users in ground communication subnets, such as various ground control towers and ground stations of ATM systems. By ATN, each end system ES carries out point-to-point communication with other end ESs in the network system. The intermediate system is the router of ATN. The routing domain of ATN network is divided into intradomain routing and interdomain routing, so an intermediate system is subdivided into Border Intermediate System (BIS) and IS. BIS is used to transfer data in interdomain subnets and IS is used to transfer data in intradomain subnets.

10.1.4 ATN and SWIM

System-wide information management (SWIM) is a core technology to share and exchange information of air traffic control in air transportations around the present world. Next Generation Air Traffic Management in the United States and Single European Sky ATM Research (SESAR) adopted SWIM to build and implement a framework for information exchange, which has proven the functionality of service-oriented architecture (SOA) to SWIM [2].

ATN system architecture works for the global deployment of ATM applications and services, while SWIM is a solution to technical defects of the ATN system architecture and the original system, which is the way of information management integrating and covering ATN from the network.

SWIM's goal is to integrate information of the virtual ATM network, and its solution is carried out system-wide rather than in a specific system or interface. SWIM, as a set of network infrastructure technology for information sharing and exchange, is expected to meet all kinds of aviation accuracy and timeliness of information interaction between users, mainly based on information and data as the center, through a loosely coupled architecture with strong openness, flexibility and robustness, that is, SOA.

Current ATN infrastructure designed based on the past communication mechanism (e.g., X.25) reaches a limited bandwidth and few manufacturers are willing to produce it. However, as SWIM adopts the standardized TCP/IP protocol, which has reached a higher bandwidth, the information sharing and collaborative decision making can be realized through a standardized interface. The Federal Aviation Administration (FAA) has begun parallel deployment of ATN and SWIM and has planned to achieve full integration of ATN and SWIM in the long term. At the same time, SESAR plans the completion of the European network deployment of IP-based ground data exchange, which lays the foundation for the full deployment of SWIM services, before 2025.

10.2 Introduction of SOA

10.2.1 Service-Oriented Architecture [3]

In one common model of application interaction, an information provider interacts with an information requester, which is based on SOA, with an information discovery process assisting.

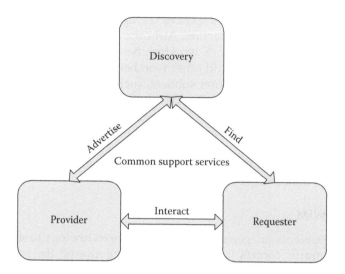

FIGURE 10.2
Service-oriented model of application interaction.

In SOA, the provider could be any service and the requester could be any consumer. The interaction between them is an easily locatable service interface. The provider offers its services through a registry where the requester can find information sources it needs and find out the location of the provider that will interact with it. The services are located with service descriptions, which is the key point of SOA (Figure 10.2).

When information providers are multiple, the requester can pick up one according to the policy or other standards. This is a form of redundancy that could be used to improve information availability.

There are no restrictions on the interaction between information provider and requester, which includes direct or mediated, push or pull, point to point or point to multipoint, etc.

In an ideal condition, information services descriptions are machine-readable, which allows the discovery and access of the dynamic information by application requesting.

10.2.2 SOA Logical Layers

Sajjad [4] proposes a seven-layer composition of the SOA architecture. According to this study, SOA was shown as a partially layered architecture that composes of services aligned with business process. It could be shown in Figure 10.3.

In this SOA architecture, the large-grained enterprise components or business-line components carry out the services, implementing the functionality and ensuring the working of the services. Combining these services to composite applications can support business process flows. In an integrated architecture, the routing, mediation, and transformation of these services, components, and processes are to be supported by enterprise service bus (ESB). The deployed services must be monitored and managed for quality of service and nonfunctional requirement.

The SOA layered architecture is described layer by layer as follows:

Layer 1: Operating system layer. This is the basic layer, consisting of legacy systems that the programs that have been developed with an outdated technology

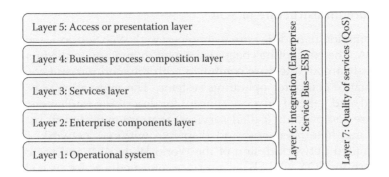

FIGURE 10.3
SOA logical layers.

make up for the vast majority of programs in many user application environments, and existing customer relationship management (CRM) and enterprise resource planning (ERP) applications, older object-oriented applications, and business intelligence applications. SOA composite-layered architecture can integrate these applications with existing systems and service-based integration techniques.

Layer 2: Enterprise components layer. This layer is responsible for realizing functionality and maintaining the QoS of the exposed services. This layer typically uses application servers to implement the component, workload management, high availability, and load balancing.

Layer 3: Services layer. The services needed for business applications are found in this layer. They are composed of dynamic or static formats or a composition of both. Therefore, the enterprise components can provide services at runtime using the functionality provided by their interfaces. The interfaces can exist independently as an exposed service description in this layer.

Layer 4: Business process composition layer. The services offered in layer 3 are defined in this layer. They are bundled into a flow to work together as a single application. These applications support special cases and business processes. IBM WebSphere Business Integration Model or Websphere Application Developer Integration Edition can be used for designing these application flows.

Layer 5: Access or presentation layer [5] explained this layer that gets more useful though being not relevant to the architecture of SOA. Still some applications seek to adopt Web Services at the application interface or presentation layer.

Layer 6: Integration (enterprise service bus [ESB]) layer. This layer provides a position independent mechanism for integration. The layer integrates services by introducing a set of reliable performance convergence such as intelligence routing, protocol mediation, and other transformation mechanisms. For instance, the Web Services Description Language (WSDL) develops binding containing addresses offering services.

Layer 7: Quality of Service (QoS). This layer is used to monitor, manage, and maintain the security, performance, and availability, which is mainly a background process with sense-and-response mechanisms and monitor tools for SOA state, including the important standards' implementation of Web Service Management (WSM).

10.2.3 ESB—Core Infrastructure of SOA

ESB is the combination of middleware technology with Web Services and the core infrastructure of SOA. The concept has been developed in analogy to the bus concept found in computer hardware architecture combined with the modular and concurrent design of high-performance computer operating systems. The motivation was to find a standard, structured, and general purpose concept for describing implementation of loosely coupled software components (called services) that are expected to be independently deployed, running, heterogeneous, and disparate within a network. ESB is also the intrinsically adopted network design of the World Wide Web and the common implementation pattern for SOA [6]. ESB, as a service intermediary, forms a chain of the service users → ESB service proxy → the service provider. The intermediary functions in various ways in different applications:

1. *Decoupling intermediary*: Customer neither knows nor cares about the actual identity of service provider, physical location, and transmission protocol and interface definition. Interactive integration code extracted outside the business logic, the central ESB platform makes central declarative definitions. ESB platform carries out protocol conversion (Web Services, http, JMS, etc.), message conversion (conversion, enrichment, and filtration), and message routing (synchronous/asynchronous, publish/subscribe, content-based routing, branching and aggregation, etc.).

2. *Service intermediary*: ESB platform provides basic services in services interaction as an intermediary. ESB platform implements SLA (reliability assurance, load balancing, flow control, cache, business control, and encrypted transmission), service management monitor (exception handling, service call and message data records, condition monitor of systems and services, and ESB configuration management), and unified security management (as it is hard to be achieved).

3. *Service orchestration*: Several services are orchestrated to form a new service. ESB supports a visualized form to define a new composite service process (workflow, BPEL, or code-level orchestration).

Therefore, the basic functions of ESB are data transmission, message protocol conversion, and routing. These three core functions are always provided by ESB in the integration of heterogeneous systems. Although SOA can also be carried out without ESB by means of SCA and BPEL, it is difficult to implement the transformation of messages protocols and dynamic routing.

Some of the original message middleware have been transferred into ESB products; such message middleware and data bus are applied more in the original EAI application integration. SOA integration is based on Web Services and regards WS as its basic management unit. Position of a service is about how to implement the business logic as a set of mutually independent self-described and interoperable entity.

SOA is concerned about the whole life cycle of service through which the business value would be achieved, while ESB is concerned about service intermediary and service integration, which is the infrastructure of SOA. SOA has two core components, one is ESB and the other is BPEL, and while ESB is infrastructure, BPEL is the service integration driven by a business process. Without SOA, ESB will lose its connected services and is only a bus with no value.

The early establishment of SOA did not contain a large and complete ESB. However, business problems should be given attention and solved using SOA. These business services will create business value, and the services can be assembled to dynamically solve changeable business requirements, making the user's business flexible and diverse. In the process of assembling, an ESB can be considered to connect these services together.

ESB requires a certain form of service-routing directory to route service requests. However, SOA may also have a separate business service catalog most basic form of which may be a design service directory used to achieve the reuse of services throughout the development of the organization. Web Services vision places an Universal Description Discovery and Integration (UDDI) directory in both business service directory and service-routing directory, thus making it possible to dynamically discover and call services. Such a directory can be considered as part of the ESB. However, the business service directory may be separated from ESB until such a solution has become widespread.

10.2.4 Industry Benefits from Implementing SOA

SOAs are built using a combination of industry standards and industry best practices. The most important aspect of an SOA is the notion of modular components. Components are interconnected through a common shared bus, often referred to as an ESB. The ESB acts as the communication platform that allows the components to communicate. Components can be written in a plug and play fashion to subscribe to the data of interest on the bus, and publish data and provide services for use by other components. The second key attribute of an SOA is the use of well-defined data formats with well-defined semantics so that the components on the bus can understand the data and use it appropriately.

In developing an SOA, each service should have a well-defined and published interface. There should be very loose coupling between clients and services and the service should be independent of the implementation technology. There should also be service levels (QoS) defined for each service. Often the services are exposed to the ESB via Web Services standards [7].

There are a number of benefits from implementing an SOA. Not only will SOA improve the code reuse that can enable lots of savings, it will also give information about business processes and how the business can be improved.

There are six benefits of SOA given in the following tables from the commercial perspective:

1. The innovation will increase the speed of loose coupling of applications.

2. Cost of software development will be reduced as loose coupling allows smaller projects to be implemented.

3. Alerts can be triggered to deal with real-time business events where help or change is needed in the organizations.

4. Unstructured processes (i.e., activities outside the scope of automation) can be integrated with the *structured* processes, which are called convergence.

5. Collaboration tools will be integrated with portals to increase productivity by exposing services of role-based user workflow.

6. It is easier for loose coupling to update applications to new rules in order to evolve the whole processes.

10.3 Research to Improve the Stability of SWIM Based on SOA

As key implementation of information exchange and data sharing in an aviation network, SWIM provides an integration of various resources from the network, including airspace management, flow management, air traffic management, surveillance management, and aircraft system to manage correspondence, GPS, surveillance data, weather, and geographical information [3]. Thus there are several varieties of application servers deployed in SWIM (Figure 10.4).

The SOA infrastructure is required to be built with great availability, allowing for operation of civil aviation, to weaken the influence of unpredictable conditions such as emergent network delay, long response time, and reduction of QoS. All countries set up SWIM-SUIT [8] based on a reliable network infrastructure.

10.3.1 General Introduction to Releasing the Thread Blocked of Application Server

The resilient SOA system is required to build SWIM for civil aviation. Blocked threads on an application server is a major problem in building a resilient SOA, as shown by a large number of practices [9].

As the main platform to deploy civil aviation business services, application servers are an important infrastructure of SWIM used to implement service logic. Obviously, the stability of application servers is extremely important for a civil aviation air traffic control system. However, numerous servers are distributed irregularly and often require thread remote calls.

The solution of the problem requires using the present framework of SOA and avoiding vast modification of the whole system. The manageability of the adjustment and ease of implementation must be ensured even involving the configuration adjustment and a small amount of refactoring SOA.

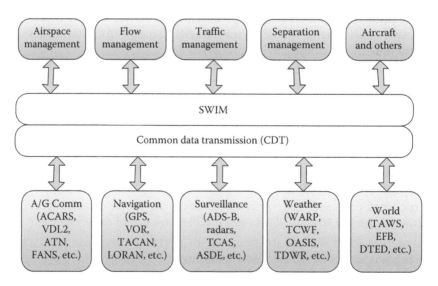

FIGURE 10.4
Architecture of SWIM based on SOA.

This section presents two methods—adjusting the physical system deployment and application servers:

1. *Adopting aggressive timer settings*: It is the key step to speed up the average service of every thread by dynamically changing timeout probability and timeout value. Queue model reveals the direct influence of timeout and average response time on system and has Remote Method Invocation (RMI) processed in 30 s before EC3-abend happens by further calculation [10]. Thus, aggressive timer settings based on the queue model will release blocked threads on some levels to strengthen stability and productivity of SOA.

2. *Optimizing the deployment of SOA services or applications*: In this method business applications are deployed with tight coupling in the chain of synchronously inter-connected components in the same physical system. Deployment of business applications that are distributed in the network and share vast synchronous communication aim to reduce RMI and improve SOA.

10.3.2 Adopting Aggressive Timer Settings: Methods Based on Self-Contained Queue Model

10.3.2.1 Application of Aggressive Timer Settings

It generally takes much longer to use a passive timer to time than the average response time. Dispatch timeout is set to 300 s while the process only takes 1–2 s. As Figure 10.5 shows, if a normal average response time of a service request is 3 s, it is still a small number of requests causing RMI timeout that interferes with the speed rate of the server seriously, even though RMI timeout value is averagely 30 s.

The coming threads are only waiting in line for deployment by server's managed tasks as the speed of the application servers is not as fast as that of coming threads. Provided that all managed tasks are blocked, servers cannot work. In this case, all programs in line will be out of time and fail. This will grandly affect stability and practicality of SOA.

In the WLM queue model, throughput of the servers will be increased as the simply quantifiable aggressive timer sets the timeout value to 10 s. However, this timer could

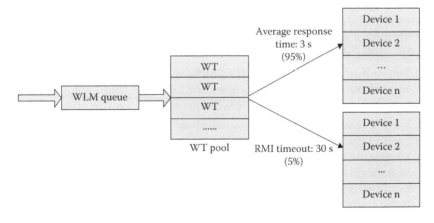

FIGURE 10.5
Influential factors of service rate.

cause the end of part of service requests and add failures. Thus, service efficiency of servers will be weakened. As a result, the timeout value of an aggressive timer should be set based on data dynamic of timeout probability in WLM queue model to speed up the service of every thread and reduce the frequency of abnormal end of EC3.

In setting the aggressive timer, its attributive timeout probability belongs to continuous data. Thus, it makes no sense to save and process data directly. In practical operation, it is an effective way that the timeout probability value can be discretized into several grades.

10.3.2.2 Implementation of Discretization of Timeout Probability Attribute of Queue Model by KDD Language Field

With respect to the convenience of calculation in algorithm design of queue scheduling, timeout probability should be processed in discretization in the first place. At present, researchers came up with various discretization algorithms for continuous data such as finite element method, finite difference method, the discretization algorithm based on clustering, etc.

The procedure of discretization method is listed as follows [11]:

1. *Attributive classification*: Confirm the quantity of required discrete values, then give the standard values of discrete radius, lower threshold, upper threshold, radius of error (ε-neighborhood), and membership degree of discrete values. Generally, these numbers can be offered by users or experts. Taking an example of timeout probability, use the language of the five discrete values: *very low, low, general, high, and very high*, and the corresponding standard samples of $a_1 = 10$, $a_2 = 30$, $a_3 = 50$, $a_4 = 70$, and $a_5 = 90$ (unit "%"), respectively. The radii of error are set to $r_1 = 2$, $r_2 = 2$, $r_3 = 2$, $r_4 = 2$, and $r_5 = 2$. The membership of discrete values can be given by the user (or expert).

2. A continuous numerical u mapping to discrete values has two problems. First, if u does not fall in the crossing range, it can be directly mapped to the corresponding discrete values. Second, if u falls within the range of the intersection of a_i and a_{i+1}, then U (nonstandard vector of u) can be calculated using a different formula (wherein a_i for the *i*th interval of standard sample points, l_i is the length of the area, A_i is the standard vector of the area, and A_a is the standard vector of the adjacent area [may be either A_{i+1} or A_{i-1}]). After that, it can calculate location according to the minimum measure values of U and A_i, A_{i+1}, and A_{i-1}, which are calculated using Hamming distance.

   The theory is precise and simple for practical calculation, not only avoiding the confusion of boundary value under the condition of a small amount of data, but also taking account of dynamic data, which is suitable for this study.

3. *Algorithm of discretization of continuous attribute based on language field theory*: The discretization method refers to defining lingual variables and their values, and which determine to the interval boundaries. With a great amount of data existing in the operating system; the database cannot load the extra data. Generally, upper and lower threshold values can be confirmed by experts as the range of attribute's value. If the count of discrete value that is defined by user is five, then a_1, a_2, a_3, a_4, and a_5 are the standard samples. As shown in Figure 10.6, using a_3 as an example,

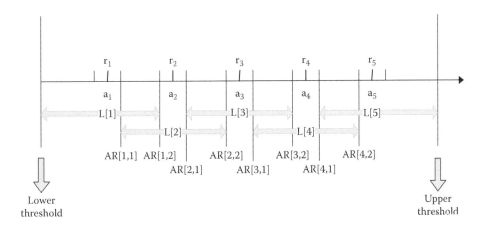

FIGURE 10.6
The range of standard sample for continuous attribute discretization algorithm.

r_3 will be the error radius of a_3. If the attribute's value is in the scope of [lower threshold, AR[1,1]], [AR[i,2], AR[i + 1,1]], [AR[4,2], and upper threshold]$(i = 1,2,3)$. It is a standard sample, beyond that will be a special sample [11].

The algorithm is illustrated as follows:

a. List is an ascending order of the attribute value table without repetition, solving AR[i,1], AR[i,2], O[i], and L[i] and setting up their membership

b. For $(i = 1; i < 5; i = i + 1)$;

c. List. Find Nearest (O[i]);

d. t = the value at the location pointed to by the pointer in the List;

e. If $t \le$ AR[i,1], then Li[i,2] = AR[i,1], List. Next; else go to g);

f. If List.EOF = true, then Li[i + 1,1] = AR[i,2] and go to i);

g. If $t \ge$ AR[i,2], then Li[i + 1,1] = AR[i,2],List. Prior; else go to i);

h. If List.BOF = true, then Li[i,2] = AR[i,1] and go to i, else solving U by interpolation formula, and determining Li[i + 1,1] and Li[i,2] by the value of Hamming distance;

i. Continue for;

j. The end.

Using this algorithm can get boundary value as Li[i,2] and Li[i + 1,1], and i = 1,2,3,4.

WLM queue model automatically adjusts the levels of timeout values of the aggressive timer referring to the previous dynamically changing data of timeout probability to weaken the influence of timeout on servers' throughput. When timeout probability increases to the error radius of critical value, timeout values can be upgraded and added to improve service. Very high timeout probability leads to servers getting blocked significantly. Servers will get rid of threads in line and release resources to get over the block as soon as possible.

4. *Analysis of the simulation experiment*: If the arrival time of WT in queue is random vector, it often follows the Poisson distribution. Each WT service time obeys expo-

nential distribution parameter to 1/EX. Among them, EX is the average service time. If set at the average service time of 3 s, the distribution parameter is a third.

There are two kinds of simulation experiments: The first situation is the variation of timeout probability when the equipment is in normal condition. The second situation is the variation of timeout probability when the equipment is out of order on some scale. The result mainly analyzes the variation of server's throughput on condition of simple aggressive timer, passive timer, and probability dynamic changes in levels of aggressive timer.

Experiment setting: If there are 40 WT pools and 20 referred-access equipment, the arrival time of WT that WLM inputs obeys the Poisson distribution, and every service time obeys exponential distribution (set EX = 3.0, distribution parameter = 1/3). The simulation experiment adopts C# (VS.NET 2010) design implementation. Random sequence using GUID as seeds makes its randomness more accurate.

Experiment 1: Variation of the throughput and timeout of WT probability on the condition that all equipment work normally. Timeout values are divided into four conditions to study according to the same group of random WT sequence: (1) timeout = 30 s, (2) timeout = 60 s, (3) timeout = 120 s, and (4) timeout = 300 s. As Figure 10.7 shows, on the condition that equipment and servers work normally, the throughput will increase with timeout values decreasing, and meanwhile the timeout probability will also increase a little. Thus, the experiment proves that setting timeout does not exert an obvious influence on server's throughput on the condition of the equipment's normal working.

Experiment 2: Part of the equipment breakdown. The breakdowns that cause timeout of WT in servers are divided into two types: one is the breakdown of the accessed equipment, the other one is the breakdown of correspondence threads (router breakdown). The experiment assumes that the breakdown occurs to 10% of equipment suddenly at the 300th second after the server starts its service and results in the timeout of all WT accessing the equipment. The number of WT queuing up in working thread pool increases due to the timeout of part of the equipment, causing that the space to use in the pool shrinks, which results in chained responses breaking down the normal procedure. Meanwhile, numerous timeouts of WT reduce the server's throughput and affects the stability of the whole SOA framework. This experiment studies four conditions according to the same group of sequence of WT from WLM: (1) passive timer setting timeout = 300 s; (2) aggressive timer setting timeout = 30 s; (3) aggressive timer setting timeout = 60 s; and (4) aggressive timer setting timeout = 120 s. The result is shown in Figure 10.8 that the average throughput decreased as the passive timer sets timeout = 300 s and the average service rate is below 1 WT/S after 20 min, which slows down the response speed of the whole server and reduces the resilience and stability of the SOA system. In the other case, the throughput remains stable with a simple aggressive timer, but timeout probability is too high to retain normal WT, which means the efficient throughput is reduced sharply.

Thus, it can be seen that WLM requires a method to keep the throughput stable and keep the timeout probability from growing too high. The experiments prove that neither simple aggressive timer nor passive timer can realize the requirement. So this chapter uses the language field theory in data mining method to set the timeout value level of application server dynamically. Its principle is based on the

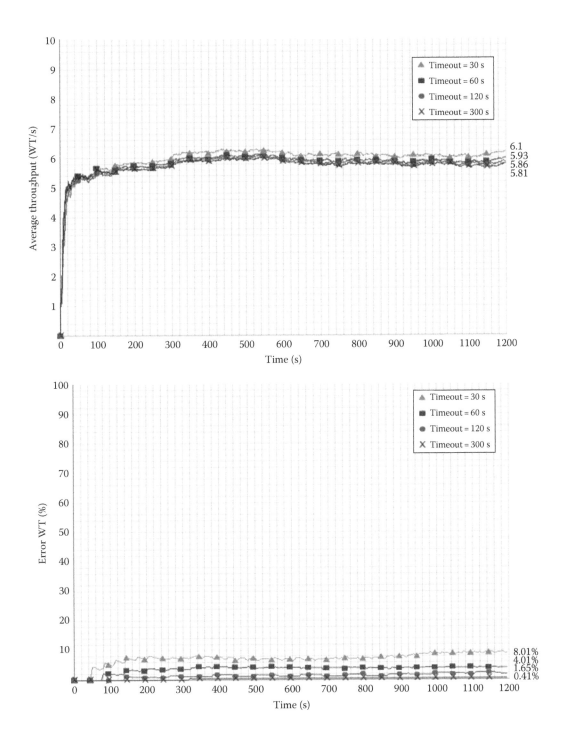

FIGURE 10.7
(See color insert.) Four kinds of throughput and timeout probability curve on the condition of trouble-free equipment.

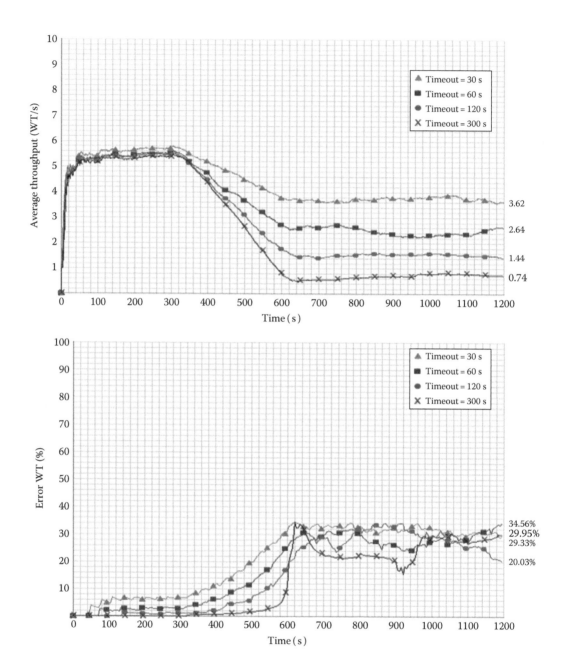

FIGURE 10.8
(See color insert.) Variation curves of 10% of equipment breakdown at the 300th second during the experiment.

timeout probability of continuous attributes that is divided into five grades: *very low*, *low*, *general*, *high*, and *very high*. The corresponding timeout probability is set to 5%, 25%, 50%, 75%, and 95%, and the language value of the membership degree is given. For example, the membership of *general* is A3 = [1.0, 0.8, 0.5, 0.2, 0] and its threshold limit is 0% to 100%, as is shown in Figure 10.9, when set r = 5%, because the timeout probability value is a continuous change, in the crossing range will

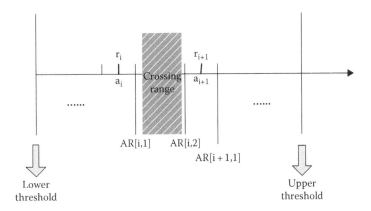

FIGURE 10.9
Crossing range.

not be able to accurately correspond to a discrete value. For example, when the probability falls within the range of cross between 10% and 20%, which can be divided into levels, we will need to use the language field theory to calculate. With the increase of the amount of data, the boundary value will be more and more accurate. This can be the continuous value of timeout probability corresponding to the discrete level of the timer. Use of a dynamic timeout value will satisfy the stability of the server throughput and try to avoid the rapid increase of timeout probability.

Experiment 3: Application server and access equipment work normally in initial state, then 10% of the equipment breaks down causing timeout of part of WT. This simulation experiment contains three conditions according to the same random sequence of WT: passive timer, simple aggressive timer, and self-contained dynamic timer. It is obvious that the throughputs of the simple aggressive timer and self-contained dynamic timer are much higher than that of the passive timer. Next, sample 10 groups of sequences of WT randomly and every group of the same sequence is simulated with the simple aggressive and self-contained dynamic timers to work out the average throughput of WT (as shown in Table 10.2).

Consequence 1: The sample means of "WT throughput" are 2.489 and 3.663. The sample variances are $S_s^2 = 0.020$ and $S_a^2 = 0.021$. The average throughput of WT of the self-contained dynamic timer is higher than that of the simple aggressive timer, and the stability of both near the sample mean is essentially the same.

TABLE 10.2

Contrast of WT Throughput (WT/S) between Adaptive Dynamic Timer and Simple Aggressive Timers

Group of Experiments	1	2	3	4	5	6	7	8	9	10
Self-contained dynamic timer	3.62	3.53	3.69	3.83	3.87	3.49	3.44	3.67	3.71	3.78
Simple aggressive timer	2.46	2.42	2.69	2.29	2.73	2.60	2.36	2.40	2.46	2.48

Consequence 2: Comparison between paired data on the same condition can use the confidence of alpha = 5% of the t test for hypothesis testing. It is proven that the throughput of self-contained dynamic timer is sharply higher than that of simple aggressive timer.

5. *The experimental conclusion*: This section presents the blocked threads of application servers in SOA and provides a method that puts forward the strategy of the dynamic aggressive timers. Timeout probability with continuous attributes and language field theory can transform discretization processing into the levels of timers. This method solves the problem of calculating the critical value of implementation. This method avoids the problem of level boundaries of chaos under the small amount of data, and does not increase the load with the large amount of data.

Although the self-contained queue model releases the blocked threads of application servers in SOA on some level, an excess of improper thread remote procedure calls still causes the block. In the next section we will focus on dynamic adjustment of application programs of SWIM in civil aviation and deployment of servers, as well as put forward methods of reducing thread remote procedure call. These methods will cooperate with the self-contained queue model to improve the resilience of SOA infrastructure.

10.3.3 Optimizing the Deployment of SOA Services or Applications

In fact, it is difficult to achieve the thread centralized application server program in a truly comprehensive way according to the features of the hardware infrastructure of civil aviation SWIM. This chapter discusses a class of short-term solutions, which can be easily applied to existing SOA framework without changing (or with just a few changes) the overall architecture. Such solutions involve only configuration adjustments and other minor optimization, and hardly require reconstruction of the SOA to carry out [9].

10.3.3.1 Optimizing and Deploying Tightly Coupled SOA Services

The basic idea of this method is to optimize the deployment of the tightly coupled network services in the SWIM or applications on physical systems. Tightly coupled SOA services refer to the relationship between multiple services called synchronously; when a blockage happens to the service the client calls, the service will not continue until it receives a response. In general, many tightly coupled SOA services often appear in a transaction, thereby forming a synchronous interconnect assembly chain in SOA.

The short-term solution is to deploy tightly coupled business applications (or services) in the synchronous interconnect assembly chain together as possible. The focus is on business applications deployed on a separate server, but sharing a lot of synchronous communication with each other (i.e., tightly coupled to each other). By redeploying to the same server, the overall condition and stability of the SOA are improved.

10.3.3.2 Advantages of Optimized Deployment

The two main advantages of tightly coupled business applications (or services) in SOA deployed together are as follows.

First, the in-process communication protocol can be taken advantage of to strengthen direct communication between applications when they are deployed in the same host.

The in-process communication can avoid all the cost remote service invocations (including serialization, encryption, traversing the network stack, and network latency), so this method can bring significant performance improvement for the entire system.

Second, it can reduce pressure on the resource requirements to deploy the tightly coupled business applications together, especially the thread-level resources. Synchronous service call through the local communication protocol in local in-process typically generates less pressure on local servers than deployment from the remote communication protocol does.

10.3.3.3 Optimized Deployment under High Workloads

Such synchronization-dependent applications deployed on a different server typically generate HTTP or RMI communications connections (Figure 10.10). For example, an external client calls application A in the server 1. Their work request will be sent to the hosted task of the server 1. However, because the communications between the applications A and B are synchronized (e.g., RMI/IIOP), the hosted tasks distributing work to application A in the server 1 must be blocked until it receives a response from the application B and when blocked, hosting tasks are all waiting in line.

If the server suffers from a heavy load at this time, only a small part of hosted task resources are available, so more and more hosted tasks are blocked on synchronous remote service calls, and the deploying ability of server 1 would be further reduced, resulting in the decline of server 1 in the rate of overall service quickly. When the service rate is lower than the arrival rate of new tasks, the server will not keep up, so the new task will be waiting in line. The worst outcome is that all hosted tasks are blocked, and the server cannot perform any work. In the case that this problem persists for a long time, queue tasks and distributed tasks will run overtime and eventually fail.

The solution to this problem is to deploy tightly coupled business applications on the same server, which apparently reduces pressure from applications on resources (Figure 10.11). Application A can use the local in-process protocol to call service B as local calls directly, which avoids occurrence of a remote call and a series of problems related to the remote delay. This will reduce the blocked managed tasks, so that the server can maintain a high rate of service, effectively avoiding the queue growth and preventing timeout of the

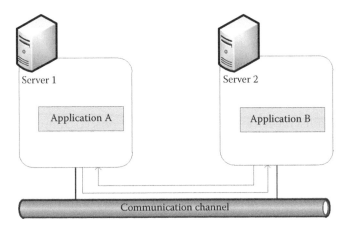

FIGURE 10.10
HTTP or RMI communications connection of a server to another server.

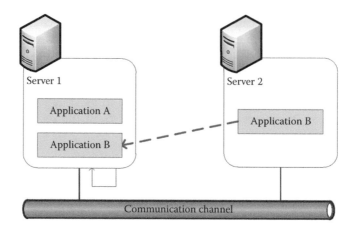

FIGURE 10.11
Deploying tightly coupled business applications on the same server.

task deployment, which in the end further reduces instability in the environment of the SOA to give a more stable and more resilient SOA system.

10.3.3.4 Preferable Service Deployment Solution Based on Clustering Analysis of Log

Due to the characteristics of civil aviation applications, the SWIM application deployment is in certain particularity, and services and applications of SWIM are also in a constant process of change. Therefore, this chapter presents a log-based cluster analysis method to find a preferable system deployment program.

The first step is data preparation—to extract attributes of service (or application) objects in log, which involves the following: (1) conform the sets of all services (or applications), (2) obtain each service server address, (3) identify the total number of times each application calls other applications, (4) record RMI for each service that occurs (including the times and RMI objects); (5) set a *moveable* for each application that cannot be migrated in order to ensure the feasibility of the optimization program, since there is a certain limitation for the migration position of applications in the civil aviation network.

The second step is service (or application) set division. A specific division method is used if the number of application programs is n, the number of servers is k to divide clusters, and the initial state of each cluster is the application set A_k of every server (excluding nonmigrated applications).

In the third step, the set will be clustered by a typical algorithm K-means, calculation of the distance between each application from the times of calls and the times of RMI occurrence.

Each cluster ($\leq N$) must meet the following two conditions: (1) comprise at least one object and (2) each object must belong to only one cluster. Then to improve the data similarity between clusters, an object is moved from one division to another based on the difference between the clusters by an iterative relocation algorithm based on the initial division. A general guideline of a preferable division is the objects in the same class are possibly *close* or similar, but the objects in different classes are *away* or different.

Suppose there are n samples of data, divided into k clusters ($k \ll n$) for t iterations, the time complexity of the serial K-means algorithm is O ($n * k * t$). Obviously, in order to reach the global optimal, division-based cluster will usually require exhausting all possible divisions, but K-means algorithm in actual use ends generally based on local optimum. The K-means dividing method that converges faster is more suitable for small and medium-sized grouping of data elements.

Usually parallel computing is recommended for solving data problems with such a large amount of computation, and K-means is a parallelized algorithm. Therefore, this chapter proposes the Hadoop cloud computing platform to solve complex computing problems in the civil aviation SWIM system. For example, in the Hadoop environment, data stored in the Hadoop distributed file system (HDFS) file is automatically divided into data blocks of 64 K. In the map phase, a data block is dealt with using a map task and the entire data file is assigned to a different node. As each node can complete i tasks and j nodes participate in the parallel computation, time complexity of parallel K-means algorithm is O (n * k * t */ (i * j)). Thus, in theory, the time efficiency of the post-parallel K-means algorithm is indeed greatly improved.

10.4 Cluster Analysis on the Hadoop Cloud Computing Platform

10.4.1 Introduction of Hadoop Cloud Computing Platform [12]

Hadoop is a cloud computing platform that can be easily developed and processes lots of data in parallel. Its main features include strong expansion capacity, low cost, high efficiency, and good reliability. First, HDFS of Hadoop uses the M/S architecture. An HDFS cluster is composed of a management node (name node) and a certain number of data nodes (data nodes). Each node can be an ordinary PC. In use, the stand-alone file system and HDFS are very similar, and HDFS can also build a directory; create, copy, and delete files; and view file contents. However, its underlying implementation is to split the file into blocks, and these blocks are stored on different nodes. Each block can also be copied and stored on different nodes in order to achieve fault tolerance. The management node is a central server that is responsible for managing name space (namespace) of the file system and access to client end on files. The data nodes in the cluster manage the data store on the node. The management node is the core of HDFS, which records how many blocks are cut by maintaining a group of data structure. The blocks can obtain from those nodes important information like the condition of each data node. Besides, MapReduce of Hadoop is an efficient distributed programming model, and a way of processing and generating a large number of data sets.

10.4.2 Exploration of Hadoop Clusters and Internet

10.4.2.1 Basic Principles of Operation of Hadoop Clusters

There are three parts of main task deployment in Hadoop: client machine, the master-, and slave nodes. The master node is responsible for the supervision of two key functional modules, such as HDFS and MapReduce. When the job tracker uses MapReduce to do parallel processing, the name node is responsible for monitoring and dispatching HDFS. The slave node is responsible for most of the machine's running and working on all data storage and command computing. Each master node is a data node but also a daemon to communicate with its master node. Daemon is affiliated with the job tracker, and data nodes are affiliated with the name nodes [13].

The client machine is assembled with all the cluster settings on Hadoop, but without master nodes and slave nodes. Instead, it is the client machine that loads data into clusters and submits description of data processing to MapReduce, and gets it back or checks out the results after work. In a small cluster (approximately 40 nodes), it often happens that a single physical device handles multiple tasks such as job tracker and name node simultaneously. As the middleware of a large cluster, a separate server is generally used to handle a single task.

There is no virtual server or management layer in the real product cluster to avoid unnecessary performance loss. Hadoop works best in Linux, operating directly with the bottom hardware facilities, which shows that Hadoop actually works directly on the virtual machine. This is superior in terms of cost, ease of learning, and speed.

10.4.2.2 Structure of the Typical Hadoop Cluster

The structure of a typical Hadoop cluster is shown in Figure 10.12. A series of racks are connected by a large number of rack switches with rack servers (not the blade servers), which is usually held out with 1 or 2 GB broadband. Ten gigabytes bandwidth is not common but can significantly increase the density of the CPU core and disk drives. The last rack switch layer connects a number of racks simultaneously together by the same bandwidth and forms clusters. Lots of servers would become the slave nodes with disk storage machines, CPU, and memory (DRAM). Also some machines would become the master nodes and these machines with a small number of disk storage machines own a faster CPU and more DRAM.

10.4.2.3 Hadoop Workflows

As business users have large amounts of data to be analyzed and processed quickly, Hadoop separates massive data and dispatches the data to the computer to do the parallel

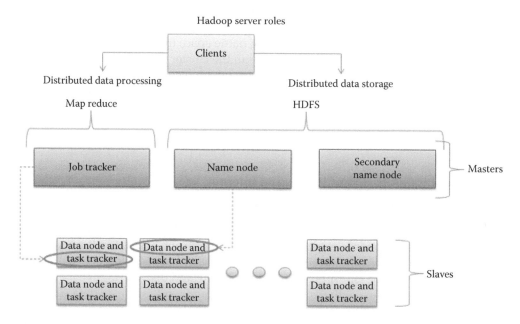

FIGURE 10.12
(See color insert.) Hadoop server roles.

process. For example, if the user has a large number of data files (such as e-mails to the customer service department): The number of times *refund* appears has to be counted in time, which is a simple task of word count. Customers load data into clusters (file.txt) and submit a description of the data analysis, - *word count*. Then clusters will store the result in a new file (results.txt), and the client reads the result document. Typical workflow is shown as follows:

- Load data into the cluster (HDFS writes)
- Analyze the data (MapReduce)
- Store results in the cluster (HDFS writes)
- Read the results from the cluster (HDFS reads)

The Hadoop cluster does not work until the data is input into it, so a large file.txt has to be loaded to the cluster above all, and the aim is fast parallel processing of data. To achieve this goal, as many machines as possible have to work simultaneously. Finally, the client will divide the data into smaller modules that are dispatched to different machines of the cluster. The smaller modules the data are divided into, the more machines can do parallel processing of data. Since these machines may be out of order, a single block of data has to be processed simultaneously on different machines in order to avoid loss of data. Each block of data would be loaded over and over again on the cluster. Generally, the default settings of Hadoop are that each block of data is repeatedly loaded three times and this feature can be set via dfs-replication parameters in *hdfs-site.xml* file.

10.4.3 MapReduce Data Flow [13]

The MapReduce client includes three working elements: input data, MapReduce program, and configuration information. Hadoop divides the job into map tasks and reduce tasks.

The job execution process employs two types of nodes: the job tracker and numerous task trackers. The function of job tracking is to coordinate all the tasks through the scheduling task running on the system. The function of task tracking is to send the progress report of the tracking task, which records the progress of each job. If the task fails, the job tracker can rearrange the task tracking.

Hadoop divides tasks into fixed size pieces in the MapReduce. These pieces are known as input splits or simply splits. Hadoop creates a map task for each split to execute each record in a user-defined mapping function.

If there are a lot of split blocks, in fact, the time of processing each split is less than that of processing the whole task. A faster machine will be able to process proportionally more splits over the course of the job than a slower machine. If we process the splits in parallel, the processing is better load-balanced if the splits are small. With a split into small particles, even if the machines are identical, failed processes or other work carried out simultaneously make load balancing desirable and the quality of the load balancing increases.

On the other hand, if splits are too small, then the total job execution time will be dominated by the overhead of managing the splits and of map task creation. For most jobs, a good split size tends to be the size of a HDFS block, which is 64 MB by default, although this can be changed or specified for the cluster (for all newly generated files) when each file is created.

Hadoop tries its best to run the map task on a node where the input data resides in HDFS, which is called the data locality optimization. We now see why the optimal split size is as small as the block size; that is it ensures the maximum size of input that can be stored on a single node. If the split was divided between two blocks, unfortunately any HDFS node can store both blocks, so some of the split would have to be transferred to the node running the map task through the network, and it is obvious that this is less efficient than running the whole map task by using local data.

A map task will output information directly to the local disk, rather than to HDFS. The map output process is an intermediate output. Its specific process is reducing the processing tasks to produce the final output, and once the work is complete the map output can be thrown away. So it cannot be stored in HDFS to avoid wastage. If a node fails before the map has been already output, Hadoop will automatically re-run the task to create a map of the output mapped on another node.

Reduce tasks do not have the advantage of data locality. That is the input to a single reduce task is normally the output from all mappers. For example, now we have a single reduce task fed by all of the map tasks. Thus, the sorted map outputs must be sent through the network to the node where the reduce task is running, they are the first to be combined and then transferred to the user-defined reduce function. Typically, the output of reduce is stored in HDFS to ensure the reliability. Therefore, the reduce output consumes network bandwidth, but only as much as a normal HDFS write pipeline.

The data flow for the general case of multiple reduce tasks is illustrated in Figure 10.13. This diagram makes it clear why the data flow between map and reduce tasks is colloquially known as *the shuffle*, as each reduce task is fed by many map tasks.

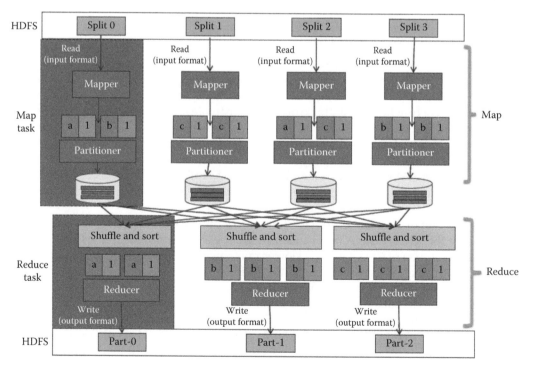

FIGURE 10.13
(See color insert.) MapReduce data flow.

10.4.4 Design and Implementation of Parallel K-Means Clustering Algorithm Based on Hadoop Platform

In the design of Hadoop parallel algorithm, the user's main job is to design and implement Map and Reduce functions, including input and output `<key, value>` key type and Map and Reduce logic, and so on [14].

As can be seen from the serial K-means algorithm, the main computational work is to assign each sample to its nearest cluster, and operations to assign different samples are mutually independent for which parallelization is required. In iteration, the algorithm performs the same operation, and the parallel K-means algorithm (Parallel K-Means), respectively, performs the same Map and Reduce operations in iteration.

Put K samples selected randomly as a focal point and save it into a file of HDFS as a global variable. Then iteration consists of three parts: Map function, Combine function, and Reduce function.

The following are the detailed descriptions of Map function, Combine function, and Reduce function.

10.4.4.1 Function Map()

The `<key, value>` pair input by Map function is the default format of the MapReduce framework, that is, *key* is offset that the current sample relates to the file starting of the input data and *value* is a character string composed of coordinate values of each dimension of the current sample. First, parse out the value from each dimension of the current sample from *value*; then calculate the distance between that and k center points and find the nearest cluster subscript; and finally, output `<key', value'>`, where the key' is the nearest cluster subscript and value' is a character string composed of coordinates of each dimension of the current sample. Function pseudocode is

```
Map (<key, value>, <key ', value'>)
{
Analyze sample object from the value, denote instance;
Auxiliary variable MinDis is initialized to the possible maximum value;
Index is initialized to -1;
For i = 0 to k - 1 do {
dis = distance between the center of the ith and instance;
if dis <minDis {
minDis = dis;
index = i;
}
}
Refer index as a key ';
Refer coordinate value of each dimension as value ';
Output <key ', value'>;
}
```

In order to reduce the communication cost and the amount of data transferred during the iterative algorithm, PK-means algorithm design Combine after the Map and merge the output data in local after Map function processing. Since the data output after each Map operation is always stored on the local node, each Combine operation is performed locally, and the communication cost is small.

10.4.4.2 Function Combine()

In the Combine function input `<key, V>` pair, *key* is the cluster (also called grouping) subscript, *V* is a character string list consisting of dimensional coordinate values of every sample assigned to the cluster whose subscript is key. First, analyze the dimensional coordinate values of each sample from the list, and corresponding coordinate values of each dimension are added together, and record the total number of samples in the list while. Key' is the cluster subscript in Output `<key ', value'>` pair; value' is a character string that includes two parts of information: the total number of samples and the accumulated coordinate values for each dimension and composition of the character strings. Function pseudocode is

```
Combine (<key, V>, <key ', value'>)
{
```

Initialize an array to store the accumulated value of the coordinates of each dimension, and the initial value of 0 of each component;
   Initialize the variable num, record the number of samples assigned to the same cluster, and the initial value is 0;

```
While (V.hasNext ()) {
```

Analyze the coordinate values of each dimension of a sample from `V.next ()` in;
The coordinate values of each dimension are added to the corresponding component in the array;

```
num ++;
}
Refer the key as a key ';
```

Construct a character string that contains num and information of all components in an array, and refer this character string as value ';

```
Output <key ', value'>;
}
```

10.4.4.3 Function Reduce()

In the `<key, V>` input of the Reduce function, *key* is the cluster subscript and *V* is the intermediate result from the transmission of each Combine function. In the Reduce function, parse out the number of samples processed from each Combine and coordinate accumulated values of each dimension of the corresponding node at first; then the corresponding accumulated values of each dimension add correspondingly in separate way, which is then divided by the total number of samples, that is, the new center point coordinate. Function pseudocode is

```
Reduce (<key, V>, <key ', value'>)
{
```

Initialize an array to store the accumulated value of the coordinates of each dimension, and the initial value of each component is 0;

Initialize variable NUM, record the total number of samples assigned to the same cluster, and the initial value is 0;

```
While (V.hasNext ()) {
```

Parse out the coordinate values of each dimension of a sample and the number of samples num from V. next ();

The coordinate values of each dimension are added to the corresponding components in the array;

```
NUM + = num;
}
```

Each component of the array is divided by NUM to get a new center coordinate;

```
Refer the key as a key ';
```

Construct a character string containing the information of coordinate values of each dimension of the new center point, and refer the character string as value ';

```
Output <key ', value'>;
}
```

Obtain coordinates of new center point according to the output of Reduce and update the coordinates to the file on the HDFS; then, perform the next iteration until the algorithm converges.

K-means algorithm is one of the most commonly used algorithms in practice, which has an absolute advantage in dealing with a large amount of data and can achieve better results. Based on the results obtained through clustering, application, or service that often call each other will be assembled in a cluster (a server), and, of course, there will be some applications frequently called by the server remotely assembled in this cluster. Apparently, it is more reasonable that these applications are deployed in the cluster (or the server), which can give advice dynamically on adjusting and optimizing application deployment.

It has great significance in solving complex computational problems to introduce a cloud computing platform into the SWIM system. This section, taking the open source Hadoop platform as an example, uses this platform for clustering computation in order to optimize deployment of SOA-based applications in SWIM in each server to further improve the stability of SOA and production availability.

10.5 Conclusion

This chapter presents the basic structure of ATN and the features of data information that includes ADS-B, an important part of a new generation of ATM systems; CPDLC, enabling pilots and controllers to carry out two-way communication directly via a data link so that they can send a variety of requests and control commands; and FIS, supporting the pilots to get all sorts of information during flight. Then, it explains the importance of the relation between SWIM and ATN. SWIM's standardized TCP/IP protocol has reached a higher bandwidth so that the information sharing and collaborative decision-making can be realized through a standardized interface. FAA plans to achieve full integration of ATN and SWIM in the long term. Also the structure and logic layers of SOA architecture are analyzed, as it consists of operating system layer, enterprise components layer, services layer, business process composition layer, access or presentation layer, integration (ESB), and QoS. As the core infrastructure of SOA, ESB is referred as a service intermediary providing functions of data transmission, message protocol conversion, and routing in the integration of heterogeneous systems. This chapter presents ESB's concept and working principle.

This chapter also focuses on the transmission control of ATN based on SOA. As the resilience of SOA has a key effect on ATN and SWIM, this chapter presents two approaches to improve network resilience based on SOA architecture:

1. *Optimize and improve the deployment of SOA services or applications*: The center of such a solution is a close coupling of business applications deployed on the same physical system as much as possible in synchronous interconnect assembly chain. It is responsible for deploying business applications dispersed in the network but shares a large amount of synchronous communication, which reduces the RMI and provides a great benefit for business application programs so as to improve the overall condition and stability of SOA.

2. *Using active timer program*: The key to solving congestion problems is dynamically changing the timeout and the timeout value to increase the probability of the average service rate for each thread and to improve the service rate for the entire server. The introduction of the queue model can intuitively show the impact of timeouts and the average response time on the system; on the other hand, further calculation can enable the system to deal with RMI within 30 seconds before EC3-abend [5] occurs. Accordingly, the timer queue based on active server model can solve the problem of congestion to some extent; the thread, thereby enhances the availability of SOA and production stability.

This chapter also introduces the Hadoop cloud computing platform while dealing with server deployment, including Hadoop common, Hadoop distributed file system, Hadoop YARN, and Hadoop MapReduce, whose main features include strong expansion capacity, low cost, high efficiency, and good reliability. And, it analyses the log of remote procedure call and the working principle of Hadoop cloud computing platform, and introduces MapReduce.

The clustering algorithm is adopted to analyze the system log and to dynamically adjust process deployment. As can be seen from the serial K-means algorithm, the main computational work is to assign each sample to its nearest cluster, and operations to assign different samples are mutually independent, for which parallelization is needed. In iteration, the algorithm performs the same operation, and the parallel K-means algorithm (Parallel K-Means) respectively performs the same Map and Reduce operations in iteration.

K-means algorithm is one of the most commonly used algorithms in practice, which has an absolute advantage in dealing with a large amount of data and can achieve better results. Based on the results obtained through clustering, applications or services that often call each other will be assembled in a cluster (a server), and, of course, there will be some applications frequently called by the server remotely assembled in this cluster.

Apparently, it is more reasonable that these applications are deployed in the cluster (or the server), which can give advice dynamically on adjusting and optimizing application deployment.

The clustering algorithm makes for increasing network flexibility and availability and reducing RMI and congestion.

References

1. ICAO DOC. Manual of technical provisions for the aeronautical telecommunication network (ATN). 97052AN/956-2001, 2001, pp. 34–35.
2. Luckenbaugh, G., Dehn, J., Rudolph, S., and Landriau, S. Service oriented architecture for the next generation air transportation system. In *2007 Integrated Communications Navigation and Surveillance Conference*, 2007, pp. 1–9.
3. Harkness, D., Taylor, M.S., Jackson, G.S. et al. An architecture for system-wide information management. In *25th Digital Avionics Systems Conference*, Portland, OR, 2006, pp. 1A6 1–1A6 13.
4. Sajjad, W. An optimum architecture for SOA. *Daffodil International University Journal of Science and Technology*, 9(1), January 2014, 47–48.
5. Travis, B. Section 3: Developing service-oriented architectures. http://msdn.microsoft.com/en-us/library/aa302164.aspx. Visited October 2014.
6. Flurry, G. Exploring the Enterprise Service Bus, Part 1: Discover how an ESB can help you meet the requirements for your SOA solution. 2008. http://www.ibm.com/developerworks/library/ar-esbpat1/. Visited December 2014.
7. Waheed, S. An optimum architecture for SOA. *Daffodil International University Journal of Science and Technology*, 9(1), January 2014.
8. Houdebert, R. and Ayral, B. Making SWIM interoperable between US and Europe. In *2010 Integrated Communications Navigation and Surveillance Conference*, Herndon, VA, 2010, pp. C4 1–C4 8.
9. Snehal, A., Alderman, R.G. Build a resilient SOA infrastructure, Part 1: Why blocking application server threads can lead to a brittle SOA[C/OL], 2007. http://www.ibm.com/developerworks/webservices/library/ws-soa-resilient/. Accessed December 2014.
10. Rica, W., Cleberson, C., Nguyen, K. et al. Problem symptoms in websphere for z/OS and their resolution. In *IBM*, 2006, pp. 20–26.
11. Zhou, Y. and B. Yang. The discretization method and its realization of continuous attribute based on language field theory. *Journal of Computer Science* (*Chinese*), 30, 2003, 63–66.
12. Apache Software Foundation. Welcome to Apache™ Hadoop®! http://hadoop.apache.org/. Visited December 2014.
13. White, T. *Hadoop: The Definitive Guide* (1st edition), O'Reilly Media, Inc., Sebastopol, CA, June 2009, pp. 75–80.
14. Venner, J. Pro Hadoop. Apress, December 2009, pp. 27–30.

Index